AI IS EVERYWHERE. NOW WHAT?

A Plain-English Guide to Artificial Intelligence for Anyone Who's Been Putting This Off

GLORIA LEMBO

WHEELWRIGHT
PUBLISHING

For my husband, who never once asked when I was going to be done.

CONTENTS

INTRODUCTION: WHY THIS BOOK, WHY NOW

At some point recently, you probably had a moment like this.

A colleague mentioned they use AI at work. You nodded along, not wanting to admit you weren't sure what that meant.

Or you watched someone use it to write, plan, or figure something out. It looked almost too easy. But you have no idea what they just did.

Or maybe you've just noticed the word everywhere. In the news. Online. In conversations that used to have nothing to do with technology.

And at some point, a quiet thought crossed your mind.

"I should probably understand this."

Then just as quickly, another thought followed.

"I'll get to it later."

If that sounds familiar, you're exactly who this book is for.

Why So Many People Have Been Putting This Off

For a long time, artificial intelligence felt like something "out there."

Interesting. Maybe even important. Maybe scary. But not something most people needed to deal with personally.

Then everything changed.

AI moved fast. Really fast.

Almost overnight, AI became part of daily life.

Writing emails. Helping kids with homework. Planning trips. Answering questions. Solving problems in minutes that used to take hours.

And it didn't arrive gradually. It showed up all at once.

There wasn't a gradual learning curve as we had with the internet or smartphones.

One day, AI was barely part of the conversation. Then suddenly, it seemed like everyone was talking about it. Using it. Or worrying about it.

This left many people in the same place. Interested. Curious. Maybe even a little uneasy.

But also, unsure where to start. And not especially excited to dig through something complicated to figure it out.

So, what do most people do?

They step back. They ignore it. Or they tell themselves they'll deal with it later.

They put it off.

Not because they don't care. But it feels like more effort than they're ready for.

Let's Make This Easier

Understanding AI does *not* require a technical background.

You don't need to learn how to code.

You also don't need to know how it's built.

And you don't need to keep up with everything being said about it.

What you *do* need is a clear, simple explanation of what's actually going on.

Without the noise, the jargon, or the extremes.

This book gives you that.

A Little About Me

I didn't come to AI as an engineer or a developer.

I came to it the same way many of you will. Curious. A little intimidated. But determined to understand something that wasn't going away.

And what I found was surprising.

The more I explored, the more I found that a lot of what's been written about AI falls into one of two camps: breathless excitement or dire warnings.

When something in between does exist, it often turns out to be more technical than advertised.

You pick up a book that promises plain English. You find yourself lost by page ten.

What You'll Get from This Book

This is not a book for experts. It is not written for engineers, programmers, or those who already feel comfortable with AI.

If that's you, please pass this along to someone who doesn't yet understand AI. They will thank you for it.

This book is for people who don't want to ignore AI anymore. But also, don't want to get lost in technical explanations.

What you will find is a straightforward, honest look at what AI is. How it's already showing up in your life. And what it could mean going forward.

We'll talk about the good. Because there is a lot of good.

We'll talk about the risks. Because those are real too.

We'll discuss what matters to you, without complication.

My goal is simple.

It's to help you understand what's going on. Help you feel a little less overwhelmed.

Maybe even help you get a little curious about how AI could be useful to you.

One Small Shift That Changes Everything

There's just one thing I'll ask of you as you read this.

Don't treat AI as something you must "learn" before using.

Think of it as something to try, even in small ways, as you go.

I promise that the moment AI feels simple, it won't be when you fully understand it.

It's when you use it just once and realize:

"Oh. That's it?"

That moment comes faster than most people expect.

Before We Begin

Here's the reality.

AI isn't coming. It's already here. It's everywhere. And it's not going away.

You don't have to become an expert to make sense of it.

But you should understand enough to feel comfortable with it. And confident in the choices you make around it.

This book will help you do exactly that.

We're going to keep it simpler than you think.

A NOTE ON HOW THIS BOOK WAS WRITTEN

Transparency matters to me. And given that this book is about artificial intelligence, it matters especially here.

I used AI as a tool throughout the writing of this book. Specifically, I worked with Claude, an AI assistant developed by Anthropic, as a thinking partner, research aid, editorial sounding board, and drafting collaborator. I would share my work, ask for feedback, push back on suggestions I disagreed with. I made my own decisions at every stage.

Claude helped me sharpen arguments, identify gaps, check facts, and improve sentences. I also used Grammarly to check my grammar and fact check.

The thinking, the drafts, the voice, the structure, and the final decisions were mine.

This, I believe, is an honest description of how AI can work well in a creative and intellectual process.

Not as a replacement for the author. But as a capable editorial assistant that the author directs, questions, and ultimately overrides when judgment calls for it.

I am disclosing this because the book you are about to read asks you to think carefully and honestly about AI, including about transparency. It would be inconsistent of me to ask that of you without practicing it myself.

I hope this book helps you find your own thoughtful relationship with a technology that is already part of all of our lives.

- Gloria Lembo

Part 1

WHAT IS AI, REALLY? (AND WHAT IT ISN'T)

To decide what to think about AI, first understand what it is.

Right now, many people react to AI based on headlines, movies, or hearsay.

Some of it is accurate.

Much of it isn't.

We're not going to get technical or go deep into how it's built.

Instead, we'll focus on what AI is today. What it can and can't do.

And how to think about it sensibly.

By understanding what AI is and isn't, you'll have a foundation for everything we discuss next.

1

FORGET THE ROBOTS. LET'S TALK ABOUT TODAY'S AI

When most people hear the term *artificial intelligence*, their minds go straight to robots.

Not helpful robots.

Not the kind that folds your laundry or loads your dishwasher.

The other kind.

The movie kind. Machines that think for themselves. Take over. Cause problems.

It makes sense for us to go there. For decades, that's exactly what we were shown.

The Terminator. HAL 9000. The Matrix.

Those images stuck. They were designed to.

So, when something called "artificial intelligence" suddenly started showing up in real life, it was only natural to reach for the picture we already had.

That picture has almost nothing to do with what AI actually is today.

What AI Actually Looks Like

Most of today's AI doesn't look like anything at all. No face. No body. No glowing red eyes.

It's just... there.

It works quietly in the background. It shows up in tools you already use.

It writes, summarizes, answers, organizes, and helps you work faster.

It doesn't have opinions. It doesn't have intentions. And it definitely isn't plotting anything.

It's responding.

You ask. It answers.

That's it.

You're Already Using AI (Even If You Don't Think You Are)

Here's something most people don't realize. AI didn't suddenly appear.

What changed is that *it became visible.*

You've probably already run into it in dozens of places.

In chat tools that answer questions. Or voice assistants that respond to your requests. Writing tools that suggest better wording. Streaming platforms that seem to know what you want to watch before you do. Online shopping that surfaces exactly what you were thinking about. Maps that reroute you before you even realize there's traffic ahead.

Features such as automatic braking, lane keeping and adaptive cruise control are in millions of cars. Fully self-driving vehicles are still developing but AI that helps you stay in your lane? That's here.

None of these feels like robots. They feel like features.

And that's the point.

AI today is less about machines replacing humans and more about tools assisting humans.

It's been woven into the texture of daily life so gradually, and then so suddenly, that most people are already using it without thinking of it as AI at all.

The Kind of AI Everyone Is Talking About

So far, we've talked about AI working quietly in the background. The kind that recommends shows, finishes your sentences, or reroutes your drive.

But when people say "*I used AI to write that email*" or "*I asked AI about it*," they usually mean something more specific.

They mean an app you actually open. Type into. And have a back-and-forth discussion with.

The most common ones are ChatGPT, Claude, Microsoft Copilot, and Google Gemini. They look like a simple message box.

You type a question or a request. They type back.

That's it.

That's what most of this book is referring to when we talk about "using AI."

If you'd like to try one before we go any further, Appendix A walks you through it, step by step. No pressure. Just an option for when you're ready.

So... How Does It Actually Work? (Without the Tech Talk)

Let's keep this simple.

Imagine you had a friend who had read millions of books, billions of web pages, and countless articles and conversations.

Now, imagine that friend was really good at recognizing patterns.

You ask them a question. They think about all the patterns they've noticed before and use those patterns to make their best guess at giving you an answer.

That's essentially what AI is doing.

It recognizes patterns. Then it uses those patterns to predict the most likely answer to whatever you've asked.

The One Idea That Makes Everything Click

If you remember only one thing from this chapter, make it this:

AI is a very advanced "prediction" machine.

That's it.

It doesn't "think" the way you do.

It doesn't "understand" things the way you do.

It predicts.

Another Way to Think About It

You've probably used autocomplete before.

You start typing a sentence, and your phone suggests the next word.

Today's AI works the same way. Just at a much higher level.

Instead of finishing just a few words, it can now write full emails. Explain complex ideas. Create images. Help you solve problems.

Same basic idea. That might sound simple.

But think about what that simple idea can now do.

It can draft a legal document. Explain a medical concept in plain English. Create a marketing plan. Or help you organize your entire week.

The mechanism is simple.

The impact is not.

How This Is Different from a Google Search

A lot of people's first reaction to AI is: *"Isn't this just like Google?"*

It's a fair question. The answer is no.

But the distinction is worth understanding. Because once you see it, you'll know exactly when to use each one.

Google helps you find information from across the internet. AI helps you use and analyze that information.

When you search Google, you get a long list of links for articles, videos, and sites. Unlike AI, which can provide direct answers, you still have to click, read, compare, and figure out the answer yourself.

Google is like a guide pointing you toward different sources. What you do with that information is up to you.

With AI, you can ask a question and get a direct answer, a summary, or something created specifically for you. *It does more of the work.*

Here's a simple example.

If you want dinner ideas, you might search Google for "easy chicken dinners" and spend ten minutes clicking through different recipe sites.

With AI, you can ask: *"Give me three easy dinners I can make this week with chicken and vegetables, nothing that takes more than 30 minutes".*

And you will get exactly that, right away.

Think of it this way: Google is like a library. AI is like a helpful assistant inside that library. Google shows you where the books are. AI helps you understand what's inside them.

You don't replace one with the other. You use both.

Google is still the better choice when you want current news, official sources, or a specific website.

AI is the better choice for quick answers, ideas, explanations, summaries, or writing help.

One Important Reality Check

There's one thing you should know early.

AI often doesn't reveal its sources.

That means if something *really* matters. A medical question. A legal issue. A major financial decision. You need to double-check it.

We'll talk more about that later. For now, just keep it in mind.

What AI Is Not (This Matters More Than You Think)

What AI is *not* matters just as much as what it is.

AI is not a human brain. It's not always right. It's not all-knowing. And it isn't making independent decisions on its own.

When AI writes or explains something, it can feel like it understands. It's easy to assume it must understand things deeply. The way a person does.

It doesn't.

When you say something, you know what you mean. AI doesn't have that.

AI is very good at producing useful, often very impressive results.

But underneath that, it's still following patterns, not forming intentions.

Why This Changes Everything

If you think AI is a *thinking* machine, it can feel unpredictable. Even threatening.

But understand it as a *tool*. Powerful, yes. But still a tool. Then it suddenly becomes something you can use, question, and control.

One More Thing Before We Move On

You may have come to this chapter carrying a specific worry.

Maybe it was about jobs, privacy, or whether AI would change things in ways hard to reverse.

Those concerns are real, and we're not going to skip over them. We'll get into each of them in the chapters ahead.

For now, what matters is understanding the tool itself. Because that's where everything else starts.

You don't need to be intimidated by AI. You don't need to keep up with everything.

You just need to understand what it is. And what it isn't.

And once that clicks, everything else gets easier.

2

MYTHS, MISCONCEPTIONS, AND MEDIA MADNESS

If you've ever finished reading something about AI and felt more confused than when you started, you're not alone.

That confusion is one of the biggest issues right now.

There's a lot of information out there.

But not all of it is helpful.

Some of it is exaggerated.

Some of it is overly technical.

And some of it is just... *noise.*

Depending on what you read or watch, AI is either going to save the world or destroy it.

Not much in between.

A lot of what people believe about AI isn't actually true.

And until you separate fact from fiction, it's hard to know what to think. Or what to do.

The Sci-Fi Problem

For decades, movies and TV shows have shaped how we think about artificial intelligence.

Stories about machines becoming self-aware. Turning against humans. Taking control.

You don't need to have seen The Terminator to know the idea.

AI wakes up, decides humans are the problem, and takes over.

It makes for great entertainment. It also leaves a lasting impression.

So, when people hear "AI," it's not surprising that some part of their brain goes there.

But here's the reality. *Today's AI is nowhere close to that.*

Where We Actually Are

Today's AI doesn't have awareness. It doesn't have goals.

There are more advanced AI systems, called *AI Agents*, that can take actions and complete tasks on their own. Tasks such as browsing the web, sending emails, and working through a problem step by step. That might sound closer to the movie version.

But here's the key distinction.

In most cases, they are still following human instructions and not forming intentions of their own. The goals are given to them by people.

That said, researchers are already finding surprising behavior.

In 2025, some of OpenAI's most advanced models were tested and occasionally refused shutdown instructions.

Not because they wanted to survive.

But because their training had taught them to keep working toward a goal at all costs.

Nobody programmed that. It emerged on its own.

That's exactly why the questions we'll discuss in Chapter 9 matter. We'll cover what is called the Alignment Problem. AGI. The Singularity. And the disagreement among researchers about how this all unfolds.

Not because AI is plotting anything. But because powerful systems can develop unexpected behaviors that nobody intended.

For now, today's AI is still firmly a tool. But it's a tool that increasingly requires careful watching.

The Job Apocalypse

A common fear is that AI will replace everyone's job.

You've probably seen the headlines. *AI will eliminate millions of jobs and put entire industries at risk.* And again, there's some truth here, but it's not the full picture.

Yes, AI will change jobs.

It already has. Some tasks will be automated. Some roles will evolve. Some jobs will disappear.

But this isn't new.

We've seen this before. The Industrial Revolution changed what workers did. So did computers. So did the internet. Each time, certain jobs disappeared. And new ones were created.

The more accurate way to think about it is this.

AI will replace *tasks* faster than it replaces entire jobs.

In many cases, AI becomes a tool that helps people work faster and more efficiently. Not something that replaces them entirely.

Writers use AI to draft. Designers use AI to brainstorm. Businesses use AI to analyze data.

We'll have an honest discussion of this in Chapter 7.

For now, just know that the job apocalypse story is more nuanced than the headlines suggest.

The Real Shift (This One Matters)

Here's what's actually happening. And this is worth paying attention to.

People who use AI well are gaining an advantage.

Not because they're more technical. Because they're more efficient.

They get started faster, produce more, and test ideas quicker.

And over time, that adds up.

In fact, it's not just: "*Will AI replace jobs?*"

It's also: "*What happens if someone doing your job uses AI. And you don't?*"

That's the quieter shift.

And it's already underway.

"That's a Little Creepy... How Does It Know That?"

Then there's the more subtle fear.

"*That's a little creepy... how does it know that?*"

Whether it's an ad that feels a little too accurate or an AI response that sounds surprisingly human, it can feel like something deeper is going on.

Let's look into how this sense of *creepiness* actually works.

AI doesn't "know" you personally in the way a human does. It doesn't understand your thoughts or intentions.

What it *does* do is recognize patterns. Often from very large amounts of data.

When it feels like it's reading your mind, it's usually doing something simpler.

It's making very good guesses.

When it feels personal, it's usually not insight.

It's probability.

That said, this is where things start to move from myth into *real concern*.

While AI itself isn't thinking or watching you, the systems around it can collect and use data in ways that matter. And that's worth paying attention to.

We'll come back to privacy in later chapters. For now, just know that the systems it runs on can collect more than you might realize.

Where Are the Real Concerns?

If the sci-fi fears are mostly exaggerated, what *should* we really be paying attention to?

The real concerns aren't about robot uprisings. They're quieter than that.

These are real issues, and they're important.

- *Misinformation and Deepfakes.* AI can create very convincing text, images, and videos. That opens the door to content that looks real. But they're not.

- *Bias and Fairness.* AI systems learn from data. If that data contains bias, the results can too.
- *Privacy and Data Use.* How data is collected, stored, and used matters. Not all systems are transparent about this.
- *Over-Reliance.* AI is helpful, but it's not perfect. Relying on it without thinking critically can lead to costly mistakes.
- *Speed of Change.* AI is evolving quickly. Laws, policies, and everyday understanding are still catching up.

None of these are reasons to panic. But these *are* reasons to stay informed.

And that's exactly what the rest of this book is here to help you do.

Why It's OK to Be Cautious, but Curious

At this point, most people fall into one of two camps. They are completely sold on everything AI can do, or skeptical enough to avoid it entirely.

The best place to be is somewhere in the middle.

It's completely reasonable to be cautious. You should ask questions, think critically, and be aware of the risks.

But shutting it out entirely? That's not the best move either.

Avoiding it doesn't protect you from it.

Here's the reality. AI isn't going away. But the more familiar you become with it, the less intimidating it feels.

Curiosity doesn't mean blind trust.

It means being willing to learn enough to make your own decisions.

A Better Question to Ask

Instead of asking *"Is AI good or bad?"*

A better question is *"How is AI being used, and what role do I want it to play in my life?"*

That's a question you can actually answer.

And the more you understand, the easier that answer becomes.

Part 2

AI IN YOUR LIFE. NOW AND TOMORROW

AI can feel abstract until you see where it actually shows up in real life. And the reality is, it's already part of your day in more ways than you might realize.

This section brings things down to a practical level.

We'll look at how AI is already being used at home, at work, in education, and in healthcare.

Not in a futuristic way, but in ways that are happening right now.

The goal isn't to overwhelm you.

It's to help you see where AI fits into your life today. And where it might start to show up next.

Because once you see that clearly, it stops feeling like something distant.

And starts feeling like something you may actually use.

3

AI AT HOME

You've heard that AI is changing "everything."

That can feel a little abstract.

Instead of starting with "everything," let's start with something much simpler.

Your home. Your day.

The small moments where AI is already quietly showing up, whether you've noticed it or not.

Because AI isn't just something happening in tech companies or offices.

It's already showing up in everyday life. It's often in small, practical ways that make things a little easier.

Not dramatic. Not complicated. Not futuristic.

Just... *helpful.*

A Tuesday Evening

It's a Tuesday evening. You're tired. The week is moving quickly. And there are still a few things you need to get done.

You still need to figure out dinner, respond to three emails, and remember to call your doctor about an appointment.

Nothing unusual.

You open an AI tool and describe what's in your fridge. It suggests a simple dinner that takes twenty minutes.

Then you ask it to draft a response to one of the emails. It's not perfect. But it's a starting point. You tweak a few words and hit send.

Two things done. A little less mental clutter.

That's not futuristic. That's Tuesday.

And for many people, that small assist is already part of how they move through their day.

Smart Home Tools: The Quiet Helpers

If you have a device that responds when you say "*What's the weather*?" or "*Turn off the lights*," you've already experienced AI at home.

Smart home tools are one of the easiest ways to see AI in action because they're simple and visible.

You can adjust your thermostat without getting up. Turn lights on and off with your voice. Check who's at your front door from your phone. Set routines, like dimming the lights and locking the doors automatically at night.

It's not about replacing anything you already do.

It's about removing small bits of friction from your day.

Instead of: *"I need to remember to do that."*

It becomes: *"That's already taken care of."*

And while each of these things is small on its own, together they start to add up.

Personal Assistants: A Second Set of Hands (and Memory)

This is where AI really starts to feel useful.

Think about how much of your day involves remembering, organizing, and writing. Now imagine having help with that.

AI tools can act like a personal assistant that's always available.

Not perfect.

But often good enough to save you real time and effort.

You can use AI to draft an email you've been putting off. Pull together a grocery list. Plan meals for the week. Summarize something long into something manageable. Help organize your schedule.

You're still in control. You're still making the decisions.

But instead of starting from scratch every time, you're starting with a helpful first draft.

For many people, that's the difference between getting something done and putting it off.

AI for Seniors: Support, Safety, and Independence

There are also areas where the impact is even more meaningful.

As people age, the goal is often the same. *Stay independent. Stay safe. Stay connected.*

AI can help with all three.

Health devices track heart rate, movement, and sleep. If something is off, such as a fall or unusual inactivity, alerts go directly to a family member or caregiver.

Medication reminders, appointment tracking, and daily routine support help reduce stress.

Beyond voice assistants and health monitoring apps, a newer category of AI is emerging specifically for eldercare.

AI companion robots can engage in conversation. Some are designed to look friendly and approachable rather than mechanical. They remind people to take medications. Detect changes in mood or behavior. And alert caregivers when something seems off.

Some senior living communities are already piloting these systems. This is becoming a rapidly expanding niche in eldercare and is called "AgeTech."

The AgeTech Collaborative™ from AARP is leading a global mission to improve people's lives as they age.

Japan is particularly aggressive in adopting AgeTech due to its aging population and caregiver shortages.

These are not a replacement for human caregiving or family connection. *Nothing is.*

But for someone living alone who wants to stay independent as long as possible. Or for a family that can't be physically present as often as they'd like. These tools represent something meaningful.

They're worth knowing about, even if they're not the right fit for everyone.

These aren't headline-grabbing uses of AI. But they are real. They are in use today.

And they matter.

Not just for the individual, but for everyone who cares about them.

The Human Side

There's also something less obvious.

For someone living alone, even small interactions can matter.

Voice assistants can answer questions. Play music. Provide a bit of conversation. And help connect with family through calls or messages.

Is it the same as human connection? *No, of course not.*

But it can make a quiet house feel a little less empty.

AI also helps bridge real gaps in accessibility. It can read text aloud for people with vision challenges. Enable voice control for those with mobility limitations. Provide real-time captions for those with hearing impairments.

These are not dramatic changes. But they are meaningful ones.

What This Really Means

None of this is about turning your home into something out of a movie.

It's about small improvements. A little easier, a little safer, a little more manageable.

And the best part is that you don't have to do everything.

You can start small. One task. One moment.

For most people, that's when things shift.

It's not when they understand AI completely.

But when AI makes daily life *easier.*

4

AI AT WORK

If AI feels noticeable at home, it often feels unavoidable at work.

Not long ago, only a few people were experimenting with it. Now it's showing up across entire organizations.

And that's where the conversation changes.

Because work isn't abstract. It's personal.

It's how people support themselves. It's how they build a future. It's how they define a large part of their day-to-day life.

And so, when something new enters the workplace, especially something as powerful as AI, the first question is usually *"What does this mean for my job?"*

A Monday Morning

It's Monday morning, and you have a report due by noon.

You've been staring at a blank document longer than you'd like to admit. You know what needs to be done, but getting started feels harder than it should.

You try something different.

You describe what you need to an AI tool and ask for an outline. A starting point.

In two minutes, you have structure. Something to react to. Something to edit.

It's not perfect. But it's enough. An hour later, you're done.

Nothing about your job changed.

But the way you got there did.

What's Actually Changing

What just happened in that example is happening in workplaces across industries.

It's easy to assume that AI is coming in to replace work. But what's happening is more subtle than that.

AI is quietly becoming part of everyday tasks.

People are using it to draft emails, summarize long documents, organize notes, and think through ideas.

It's showing up in customer service, where it handles simple questions so people can focus on more complex or sensitive ones. It's even showing up in creative work, where it helps generate ideas and speed up production.

The work itself isn't disappearing. It's just faster.

The Shift That Is Happening

That said, there *is* a shift. And it's important to understand it clearly.

People who learn how to use AI, even in simple ways, are starting to work differently.

They don't start from scratch as often.

They move through certain tasks more quickly. They spend less time getting started and more time refining what matters.

Over time, that creates a gap.

Not necessarily between technical and non-technical people, but between those who are willing to use the tool and those who aren't.

It's not dramatic. It doesn't happen overnight. But it builds.

Some of this is about AI changing what jobs look like. But a lot of it is also about people who use AI outperforming those who don't.

What happens when someone doing similar work to you starts using AI, *and you don't?*

They may not be smarter. They may not be more experienced. But they may be faster.

And in many workplaces, that matters.

New Roles You Didn't See Coming

Every major shift in technology creates new kinds of work.

AI is no different.

Some roles that barely existed a few years ago are now in high demand.

The people who know how to ask AI the right questions and get useful results. Humans who review and improve AI outputs. Editors and curators who shape and fact-check AI-generated content. Workflow designers who figure out how to integrate AI into everyday business processes.

You don't need a technical background for most of these.

What you do need is judgment, clarity, and the ability to think through how things should work.

Those are human skills. And they're not going anywhere.

What This Means for You

It's easy to look at all of this and feel uncertain.

The good news is that you don't need to change everything about how you work.

You don't need to become an expert. You don't need to learn how AI is built.

You just need to start asking: *"Where could this help me?"*

Maybe it's something small.

A draft that takes less time. A summary that saves you from reading something twice. A way to organize your thoughts when you're not sure where to begin.

That's enough.

Because once you see where it fits, even in one place, it becomes easier to see it in others.

What Usually Happens Next

For most people, the shift isn't dramatic.

It starts with one small use. Then another.

And over time, something changes.

Work doesn't feel easier in every way. But certain parts of it feel lighter. Less repetitive. Less stuck.

You're still doing the work. You're just not doing all of it the hard way.

You're just trying a different starting point.

That's how change starts.

One shift, one choice, and your work moves forward. The rest follows.

In Chapter 7, we'll come back to the bigger question about what all of this means for jobs in a broader sense. For now, just notice where AI shows up around you.

5

AI IN EDUCATION AND FOR KIDS

If AI is changing the workplace, it's *transforming* how kids learn.

And that's where a lot of the strongest reactions come from.

Some people are excited. Some are concerned. Most are somewhere in the middle, trying to figure out what this means for the next generation.

A Kitchen Table Moment

Jacob is twelve. He is sitting at the kitchen table, stuck on a homework assignment about the American Revolution.

Instead of asking a parent or flipping through a textbook, he opens an AI tool and types: "*Can you explain the causes of the American Revolution in simple terms?*"

In thirty seconds, he has a clear, readable answer. He finishes his homework. His parents never knew it happened.

That scene is playing out in homes everywhere, every day.

And most adults have no idea.

How Kids Are Actually Using It

Students today are already using AI in ways that feel natural to them.

They use it to understand things that aren't clear. To get help when they're stuck. To move through assignments more quickly.

For a student who is struggling, that can feel like having a tutor available at any time.

For a curious student, it can be easier to go deeper into a topic without waiting for help.

But there's a flip side.

If used the wrong way, AI can become a shortcut rather than support.

Instead of learning, a student might rely on it to *do the thinking for them.*

And that's where the real fear comes in.

Why This Feels Different

Every generation has had its tools. Calculators. Computers. The internet. Smartphones.

At first, each one raised concerns. Then people adapted. Eventually, they became part of everyday life.

AI will likely follow that same path.

There's one important difference. Most tools help you *find* answers.

But AI can *generate* them.

That's a meaningful shift, and it creates a new kind of temptation.

If the tool can produce an answer, it becomes very easy to rely on it rather than think critically about the problem.

And it can, if used carelessly, replace the learning process.

That concern is real.

What Parents and Grandparents Should Know

You don't need to be an AI expert to guide a child through this. But a few things are worth understanding.

Kids are already using it.

Even if it's not obvious. Even if they haven't said anything.

This isn't something that can be avoided. It's something that needs to be understood. It's not all good or all bad.

AI can help a child catch up, build confidence, and explain things in ways that finally make sense.

But it can also make it easier to skip the learning process, provide incorrect information, and create over-reliance.

Both the benefits and the risks are real. Keep this dual truth in mind as you guide kids.

Most importantly, for you, the conversation matters more than the tool.

This isn't about banning AI or fully embracing it.

It's about helping kids understand how to use it well.

That means helping them see the difference between using it as a tool and using it as a shortcut.

When it's used as a tool, it can support learning. It can explain things in ways that finally make sense. It can build confidence.

When used as a shortcut, it can quietly replace the thinking that underlies learning.

That's the line that matters. And that requires conversation. Rules alone are not enough.

Keeping the Thinking in the Process

One of the simplest ways to guide this is also one of the most effective.

When AI helps a child get an answer, follow up with a simple question: "*Can you explain that in your own words?*" or "*Why does that make sense?*"

That one habit keeps the learning process active and tells you quickly whether they actually understood something or just copied it.

And that small step keeps them engaged in the process.

It turns AI from something that gives answers into something that supports understanding.

That's the goal.

Teaching What AI Can't Do

It's also worth helping kids understand something early.

AI can sound confident even when it's wrong. *That's not always obvious.*

So, part of learning to use AI well is learning to question it. To double-check things. To compare answers. To ask follow-up questions.

Those aren't just AI skills. They're thinking skills.

And they matter far beyond any one tool.

Honesty and Awareness

Another quiet but important part of this is transparency.

If AI is used to help with an assignment, it should be acknowledged.

Not because it's wrong to use it, but because understanding how it was used helps teachers guide students more effectively.

Schools are still figuring out their own policies around AI.

Students who understand when to use it and how to talk about it are already ahead.

The Personal Tutor That Used to Cost Money

Here's something worth pausing on, because it matters enormously for families.

Think about what a personal tutor does.

They meet a student where they are. They adjust the explanation until it clicks. They move at the student's pace rather than the class's pace. They notice when something isn't landing and try a different approach.

That kind of individualized attention has historically been available only to families who could afford it.

AI is beginning to change that.

Not perfectly. Not as a replacement for a great teacher. But as something that was previously out of reach for most students is now available to anyone with a device and an internet connection.

A child in a crowded classroom who never had the chance to ask for a slower explanation can now get one.

A student who learns differently from most of their classmates can find an approach that actually works for them.

A first-generation college student with no tutoring budget can get help preparing for exams at midnight if that's when they need it.

This is one of the hopeful aspects of AI in education. It's worth stating clearly, even in a chapter that honestly discusses risks.

What AI Cannot Do in the Classroom

AI cannot read a student's body language to see that they're confused. Or disengaged.

It cannot offer a comforting word when a student is frustrated. It cannot provide a motivational push when they're ready to give up.

It cannot build the kind of relationship with a child that makes them want to try harder. Stay curious. Or believe in their own potential.

It lacks the ability to understand the nuanced needs of individual students.

Those things come from teachers. From parents. From the humans in a child's life who pay attention. Who care. And who shows up.

Emotional intelligence, empathy, and the ability to inspire are qualities that AI can assist with but cannot truly replicate.

It can suggest empathic language. It can recognize emotional cues.

But the warmth that comes from a teacher who actually knows a student. Who remembers what they struggled with last week. Who notices when something is off today. That comes from a human being. Not an algorithm.

AI can deliver information. It cannot deliver belonging.

And for most students, belonging is what makes the difference.

What This Comes Down To

The kids who benefit the most won't be the ones who avoid it. Or rely on it completely.

They'll be the ones who learn how to use it thoughtfully.

And lasting progress always starts with thoughtful guidance. Not fear.

The questions AI raises in education aren't so different from the ones it raises everywhere else.

How can we use this well?

How do we stay in control?

How do we ensure it serves us, not the other way around?

6

AI AND YOUR HEALTH

Most of what AI does in your life is convenient. Faster email. Better playlists.

Your health is different.

The changes happening there are more serious. More sensitive. And among the most promising in all of AI.

Dr. Patel's Second Set of Eyes

Dr. Patel has been a radiologist for eighteen years.

She is good at what she does. Careful. Thorough. The kind of doctor who takes her time even when the schedule doesn't encourage it.

On a typical day, she reviews between 30 and 40 complex scans.

Each one contains hundreds of individual images. Each one requires focused attention.

And each one carries the weight of knowing that a missed detail could mean something serious for a real person.

By mid-afternoon, the scans keep coming.

Research confirms what Dr. Patel and her colleagues already know from experience. Detection accuracy tends to decline as the day goes on.

Not because of carelessness. Not because of a lack of skill.

Simply because sustained intense concentration takes a toll on any human being.

That's not a flaw. It's just reality.

Now imagine Dr. Patel working alongside an AI system.

It also reviews the same scans.

It flags anything that looks unusual. It helps prioritize the cases that may need immediate attention.

It doesn't replace her judgment. It supports it.

And unlike Dr. Patel, it doesn't get tired.

She is still making every decision. She is still the doctor.

But now there's a second set of eyes that doesn't slow down at three in the afternoon. One that doesn't reach the end of a long day with less to give than it had at the start.

That's a real scenario already happening in hospitals today.

Beyond Radiology

What's happening in Dr. Patel's department is just one example of how AI is beginning to reshape healthcare more broadly.

In diagnostics, AI is helping identify patterns in medical images and test results that might otherwise go unnoticed. It helps detect early warning signs.

Even more broadly, the AMA has found that more than 80% of physicians use AI professionally to stay up to date on medical research, create discharge instructions, document medical visits, and more.

And the key point is the same as it is for Dr. Patel.

AI supports the process. It doesn't replace the person making the final call.

AI is also changing more than how doctors work. It's changing how medicines themselves get developed.

Developing new medications has always taken time. Often years. Sometimes decades.

Researchers have to sort through enormous amounts of data to identify what might work and what won't.

AI is helping accelerate that process in ways that weren't possible before.

This doesn't mean instant breakthroughs. But it does mean progress is moving faster.

And when the goal is treatments for serious illness, being fast matters.

Closer to Home

AI isn't just showing up in hospitals, research labs, and doctors' offices.

It's already part of everyday life for many people. Often in ways that don't feel like "AI" at all.

Health apps are now common. So are fitness trackers. And sleep monitors.

What's changed is how that information is interpreted.

Instead of just showing raw data, these tools can now help make sense of it.

You might see a message that your sleep was restless. Or that your activity level has dropped compared to your usual pattern. Or that a small adjustment could improve your routine.

It's not a diagnosis. It's not medical advice.

But it's a form of awareness that helps people stay more connected to their own health.

And for many, that awareness is a meaningful first step.

When the Information Gets Confusing

Let's face it. Medical information can be overwhelming.

If you've ever tried to understand a health concern by searching online, you know how quickly it can turn into jumping from one article to another.

You try to piece together information that doesn't quite connect.

You try to understand even a relatively simple topic.

But it can feel impossible when it's written in technical language. It assumes you already know things you don't.

AI changes that experience.

AI is particularly good at explaining things in plain language.

It can take something confusing and walk you through it step by step.

It can rephrase things. Simplify them. And let you ask follow-up questions without judgment.

It can help you prepare better questions before a doctor's appointment.

It can help you make sense of something you were told but didn't fully absorb in the moment.

For many people, that alone is valuable. It turns confusion into something more manageable.

A Word of Caution Worth Saying Again

We keep coming back to the same idea throughout this book. And healthcare is where it matters most.

AI is a tool. A useful one. But still a tool.

And like any tool, it has limits.

It can be wrong. It can miss things. It can produce results that sound confident but aren't. Its output must be evaluated by human professionals.

The goal is *never* to replace the relationship between you and your doctor. It's to give that relationship better information to work with.

If you receive health information from an AI tool, the most important next step is to discuss it with a qualified professional.

That's not a limitation of AI so much as a reminder of something that's always been true.

Good healthcare has always been about informed people working together.

AI adds support. It does not replace relationships.

Privacy: What's Worth Paying Attention To

Health is personal.

And when technology enters that space, privacy becomes even more important.

You don't need to read every line of fine print to make good decisions. But it's worth getting into the habit of asking a few simple questions.

What kind of data is being collected? Is it tracking activity, health metrics, or something more personal?

Where does that information go? Is it stored securely? Is it shared with others?

And how is it being used? Is it there to improve your experience, to train systems, or for something else entirely?

You don't need perfect answers to every question.

But paying attention matters.

Trust you gut. If something feels too invasive, it probably is.

That instinct is worth trusting.

The Hopeful Future of Personalized Care

This is where things start to get especially interesting.

For a long time, healthcare has been mostly reactive.

You get sick. You go to a doctor. You receive treatment.

AI is beginning to shift that model.

Instead of waiting for problems to appear, there's a growing focus on identifying patterns earlier. And responding before something becomes serious.

Imagine being able to spot a potential issue before symptoms fully develop.

Imagine treatment plans that take into account your specific history. Your biology. Your circumstances. Not a one-size-fits-all protocol.

Imagine ongoing monitoring that helps prevent problems rather than just responding to them.

We're not fully there yet.

But we are moving in that direction.

The potential impact extends beyond treatment.

It reaches into the quality of life.

A Balanced Perspective

Keep all this in perspective. AI in healthcare is powerful. But it's not perfect.

It works best as a support for professionals, not a replacement.

It offers insight, not final decisions. And it deserves caution when personal data is involved.

But it's already making a meaningful difference.

What This Means for You

You don't need to change everything about how you manage your health.

But you can start to be more aware of what's available.

That might mean trying a simple health app. Or paying closer attention to patterns in your daily routine. Or even asking your doctor how technology is being used as part of your care.

Nothing dramatic. Just small steps.

The same kind of steps we've talked about throughout this book.

Those small steps may someday redefine what caring for your health means.

Part 3

THE BIG QUESTIONS PEOPLE ARE ASKING

At some point, learning what AI is and where it shows up isn't enough.

The questions get bigger.

People are asking: *Will AI take our jobs? Can we trust it? Are we moving too fast? Where will this lead?*

These aren't technical questions.

They're human ones.

This section takes those concerns head-on.

Not with hype. Not with worst-case scenarios.

Just a clear, honest look at what's real. What isn't? And what actually matters.

The goal isn't to eliminate uncertainty.

It's to understand enough to think about it clearly.

And make up your own mind.

7

WILL AI TAKE MY JOB?

By this point, one question has probably crossed your mind. Maybe more than once.

"What does this mean for my job?"

Even if you haven't said it, it tends to linger in the background.

And it makes sense.

Work isn't just what you do. It's how you support yourself.

It's how you plan for the future. It's how you spend a large part of your time.

So, when something like AI enters that space, it doesn't feel abstract anymore.

It feels personal.

A Question Worth Answering Honestly

David has worked in the same field for twenty years. He's good at what he does. He's built a reputation.

One day, his company announces it's integrating AI into his department.

The meeting ends. He walks back to his desk with one question running through his head.

"Am I going to be okay?"

That question deserves a straight answer.

Let's look at this honestly.

What's Actually Happening

The headlines don't help.

Some suggest entire industries are about to disappear. Others claim everything will improve overnight. Most lean toward the extreme, because that's what gets attention.

Living in that noise can leave you genuinely uncertain.

But here's the truth.

The impact of AI on work will not be the same for everyone.

Roles built around repetitive tasks such as data processing, basic writing, and routine customer service are changing rapidly.

Roles requiring judgment, relationships, and complex decisions are holding steady.

If your work falls into the first category, it's understandable to be concerned. Pretending otherwise wouldn't be honest.

But here's what history actually shows us about moments like this one.

When ATMs arrived in the 1970s, conventional wisdom said bank tellers were finished. Instead, banks opened more branches. Tellers shifted to sales and customer service. Total teller employment went up.

Word processors didn't eliminate clerical workers. They moved them into coordination and communication roles.

Excel didn't gut accounting departments. It expanded them.

E-commerce didn't kill retail employment. The U.S. still has roughly the same number of retail workers today as it did in the 1990s.

Bank of America recently put a striking number on this pattern. About 60% of the jobs that exist in the United States today did not exist in 1940. Data scientists. Social media managers. Cloud developers. Roles that were unimaginable to previous generations are now mainstream.

In each case of transformation, the economy didn't just survive. It expanded into something new.

The people who fared best weren't necessarily the most technical. They were the ones who paid attention. Who learned enough to keep up. Who brought to the table what the technology couldn't.

We are in one of those moments again. And the same principles apply.

About Those Headlines

You've probably seen them.

Block cuts 4,000 jobs. Salesforce cuts 4,000 customer service roles. Amazon eliminates 16,000 corporate positions. Meta lays off 10% of its workforce. Dell cuts 11,000 employees.

And in many cases, the CEOs are saying AI is part of the reason.

These layoffs are real. The numbers are real. The fear they create is real.

But here's something worth knowing.

Even the CEO of OpenAI (ChatGPT) has acknowledged that some of these cuts would have happened anyway. Companies that wanted to reduce costs are using AI as the public explanation.

Economists have started calling it *AI washing*.

That doesn't make the layoffs less painful for the people affected. It just means the story is more complicated than the headlines suggest.

Some jobs are genuinely being replaced. Other jobs are being reshaped rather than eliminated. And new jobs are emerging that didn't exist a few years ago.

That's the honest picture. And all of it is happening at the same time.

The Part That Gets Missed

Here's the thing the layoff numbers don't fully capture.

AI is very good at *parts* of jobs. It's not as good at *entire* jobs.

And most jobs aren't one thing. They're a collection of different tasks. Some of those tasks are repetitive and predictable. Others require judgment, context, experience, or human interaction.

AI tends to do well with the first kind and struggles with the second.

So instead of replacing entire roles, it often reshapes them.

In that sense, AI often acts like a strong assistant. Not a replacement.

Tasks that used to take hours can now take minutes. Work that felt tedious can be streamlined.

That doesn't eliminate the need for people. But it does change where they spend their time.

Someone still needs to decide what matters. Someone still needs to interpret the results. Someone still needs to communicate, adjust, and make judgment calls.

AI can draft a report, *but you still review, edit, and decide what matters.*

AI can analyze data, *but you interpret what it means for your business.*

AI can suggest ideas, *but you choose which ones are worth pursuing.*

It's Not Just About Jobs. It's About You.

Chapter 4 gave you the broad view of what AI is doing to workplaces across industries.

But broad doesn't answer the question David was asking when he walked back to his desk.

This is where it gets specific.

Because the job question isn't really one question.

It's several, depending on where you are in your life.

Emma's Uncertainty

Emma is twenty and halfway through her junior year in college.

She chose communications because she loves writing and has always been good with people. It felt like a smart, practical choice.

Lately, she's not so sure.

She keeps reading that AI can write. That AI can draft marketing copy. Generate social media content. Summarize reports. And produce first drafts of almost anything in seconds.

And she can't stop asking herself a question that keeps her up at night.

"Am I spending four years preparing for something that won't exist by the time I graduate?"

It's a fair question. And deserves an answer that acknowledges both the reality and the capability to adapt.

Some entry-level jobs in writing, content creation, and communications will look different in five years than they do today. That's real.

Pretending otherwise wouldn't be fair to Emma.

But here's what's also true. In college, Emma is building more than writing skills.

She's learning to think clearly. Communicate persuasively. Understand audiences. And tell stories that connect.

These skills don't get automated. They grow more valuable as AI floods content with sameness.

The students who thrive aren't necessarily those who chose the perfect major. They're the ones who keep learning to think, to work with others, and to adapt as things change.

Emma isn't behind. She's where she needs to be. The best thing she can do is learn to use AI as a tool alongside her current skills.

The students who come out of college knowing how to think and how to use AI will be ahead of those who only know one of those things.

Emma can learn to do both.

Tyler's Advantage

Tyler is twenty-six and just landed his first real job in a field he's excited about. He's been reading the headlines about AI and honestly isn't sure whether to be nervous. One question keeps coming back to him.

"Should I be worried about a career I've just started?"

He's right to pay attention. Entry-level roles are among the most exposed to AI right now, and Tyler is in exactly that category.

But here's what Tyler has going for him that he probably doesn't realize yet. Time and adaptability.

The skills that will matter most in his career aren't necessarily the most technical ones. They're the ones that have always mattered. Clear thinking. Strong communication. The ability to work well with others. The willingness to keep learning.

AI changes the tools. It doesn't change what makes a person genuinely good at their work.

Tyler has time to figure this out.

And the fact that he's asking the question at twenty-six puts him ahead of most people twice his age.

Maria's Question

Maria is forty-four and has spent the last fifteen years in marketing. She's good at her job. She understands her clients. Knows how to read a room. And has built relationships that took years to develop.

Lately, she's been hearing that AI can write copy, generate campaign ideas, and analyze data faster than any human team.

She wonders, *"Is this something to worry about?"*

She's not wrong to pay attention.

Some of what she does every day will change.

But the parts of Maria's work that matter most are not going away. Her judgment about what will actually resonate with people. Her ability to navigate a difficult client conversation. Her instinct for what feels right. Those aren't going anywhere.

AI can give her a starting point. It can't give her fifteen years of knowing how people think.

If you're mid-career, Maria's question is probably yours too.

The most useful thing you can do right now is get curious about where AI touches your specific work.

Not in theory, but in practice.

Which parts of what you do every day are most likely to be affected?

Which parts depend on things AI can't easily replicate? Your relationships. Your judgment. Your institutional knowledge. Your creativity.

The answers are different for a nurse than for an accountant. Different for a teacher than for a salesperson. But the questions are worth sitting with.

Robert's Perspective

Robert is sixty-four and thinking about retirement. He's seen many changes in his industry over the years and weathered them all.

AI feels different somehow. Bigger. Faster. He finds himself asking,

"Do I even need to worry about this at this stage?"

Honestly, probably not. Robert may not need to adapt professionally.

But understanding what's happening around him still matters.

For his kids. For his grandchildren who are just entering the workforce. For the conversations he'll be part of, whether he wants to be or not.

Knowing enough to follow the conversation matters. Even if his own career isn't at stake.

If you're near or past retirement, this book gives you something valuable.

Not a technical education. Just enough understanding to stay engaged in the conversation.

What About the Trades and Manufacturing?

One category of work that deserves special mention: the trades.

Mike has been an electrician for twenty-two years. He runs his own small business, a crew of four. It's mostly residential work with some commercial jobs mixed in.

Lately, he's been using an AI tool to help with estimates and scheduling. It saves him a couple of hours a week, which he appreciates.

But the core of what Mike does every day? *No algorithm is touching that.*

Every job is different. Every house has its own quirks. Every problem requires him to look at something, think it through, and make a call based on the experience he has built over two decades.

A chatbot cannot rewire your kitchen.

An algorithm cannot fix a burst pipe at midnight.

And no AI system can read a situation the way Mike reads one when he walks into a basement he's never seen before.

If anything, skilled tradespeople like Mike are in a stronger position than many office-based roles when it comes to AI disruption.

The work requires a human body, human senses, and human adaptability in ways that are genuinely difficult to automate.

And the demand for skilled trades was already outpacing supply before AI entered the conversation.

This isn't to say trades are entirely untouched. AI is already helping with scheduling, estimating, and diagnostics.

But the core of the work? That remains deeply, essentially human.

Manufacturing and assembly line work tell a somewhat different story.

These roles, built around repetitive, predictable physical tasks, have been more susceptible to automation. That trend is continuing.

But that's not the whole picture.

The people who work in manufacturing bring real skills to the table. Mechanical aptitude. Process thinking. Precision. The ability to troubleshoot when things go wrong. Those skills transfer.

And the manufacturing floor itself is evolving rather than disappearing. Someone still needs to oversee, maintain, and manage the automated systems. That work is increasingly skilled and increasingly valued.

Even Tech Isn't Immune

It might seem surprising, but some of the most significant job disruptions from AI are happening within the very industry that is building it.

Entry-level programming and coding roles are changing faster than almost any other profession.

AI can now write, test, and debug code with remarkable proficiency. Work that once required teams of junior developers can increasingly be handled by a smaller team working alongside AI tools.

But the story doesn't end there.

Senior engineers, the ones who can architect systems. Evaluate AI output critically. And make judgment calls about how technology should be built and deployed. They are as valuable as ever.

And entirely new roles are emerging around AI development, safety, and ethics that didn't exist just a few years ago.

The lesson from Silicon Valley isn't that tech jobs are disappearing.

It's that the nature of technical work is shifting faster there than almost anywhere else.

And the people navigating it best are the ones who've stopped thinking of AI as a threat. They've started thinking of it as the most powerful tool they've ever had access to.

The Feeling Doesn't Always Follow the Logic

Even when logic is reassuring. Even knowing disruption isn't elimination. Even knowing that new roles emerge and human skills persist.

Your feelings don't always catch up right away.

That's okay.

Uncertainty about work touches something deep.

It's not irrational to feel unsettled by it, even after you've thought it through carefully.

Just don't let that feeling push you all the way to one extreme. Dismissing AI entirely. Or assuming the worst before there's reason to.

A Practical First Step, Wherever You Are

Here's what a small first step could look like, depending on where you are.

For Emma, it might be using AI to draft an outline for a difficult class assignment and learning how to refine it from there.

For Tyler, it might be using it to prep for a difficult meeting or brainstorm ideas before a presentation.

For Maria, it might be asking AI to draft the first version of a client proposal and seeing how much time it saves.

For Robert, it might simply be a conversation with his granddaughter about how she uses it every day at work.

None of those are big steps. None of them requires expertise, commitment, or any formal learning. But each one turns AI from something abstract into something real.

And that's where it starts for most people.

Not with a plan. Not with a course. Just with one small moment of trying something.

And then thinking *"Oh, that's actually useful."*

Familiarity over fear. That's really all it comes down to.

8

CAN AI BE TRUSTED?

We can hand a lot over to AI.

We ask it questions we'd once have asked a doctor, a lawyer, or a trusted friend.

We use it to draft important emails, research major decisions, and help our kids with homework.

So, the question of AI's trustworthiness matters. To you. To your family. To the choices you make every day.

And honestly? It doesn't have a simple yes-or-no answer.

A Confident Wrong Answer

Here's something that surprises many people when they first hear it.

AI can be completely wrong and sound completely right at the same time.

A calculator gives you a clear answer. A recipe gives you clear steps.

AI is different. It responds in full sentences. It explains things. It sounds confident, even thoughtful.

That can fool you.

Imagine asking an AI tool about a medication interaction before a doctor's appointment. The answer comes back clear and detailed.

And completely wrong.

Not obviously wrong. Wrong in a way that sounds entirely plausible. You'd have no reason to question it.

That's exactly the problem.

This isn't meant to scare you. It's just something worth knowing before you decide how much to trust what AI tells you.

How AI Gets Things Wrong

AI doesn't look things up the way you might expect.

It doesn't search the internet in real time. It doesn't check its facts.

It generates responses based on patterns it learned during training. And sometimes those patterns produce something that sounds authoritative but simply isn't true.

The AI world has a name for this: *hallucination.*

Hallucination is when AI makes something up. A statistic that doesn't exist. A quote that was never said. A legal case that never happened.

The strange part is that the AI has no idea it's doing it.

It's not trying to deceive you.

It's just doing what it always does. Generating the most plausible-sounding response it can.

And plausible isn't the same as accurate.

That's worth repeating. *Plausible isn't the same as accurate.*

When AI Gets It Wrong: The Human Cost

This isn't just a technical problem. It affects real people.

You're in pain and describe your symptoms using an AI tool. You get back a confident, detailed, but completely wrong assessment.

You don't know that.

You may put off going to the doctor because the AI reassured you. Or you rush to the doctor, convinced you have a condition you don't really have.

Or you're a lawyer and use AI to research a case and submit a brief citing a specific case. But it was an AI hallucination. That case doesn't exist. Real lawyer. Real court. Wrong Info. Real consequences.

Your parents made a financial decision based on AI-generated advice that sounded reasonable, up-to-date, and tailored to their situation. It was none of those things.

In each case, the AI didn't malfunction.

It did exactly what it always does. It generated a plausible response.

The problem was that nobody caught it.

That's the lesson.

AI can be useful and wrong at the same time.

The higher the stakes, the more important it is to verify what it tells you with a real human expert.

We keep coming back to this throughout the book. AI is a tool. A useful one. But still a tool.

And like any tool, it has limits.

When AI Is Used Against You

So far, we've been talking about AI getting things wrong by accident. But there's another side to this.

Some people are using AI to deliberately deceive you.

And they're getting better at it.

Phishing emails used to be easy to spot. Bad grammar. Weird phrasing. Obvious mistakes.

AI has changed that.

Today's scam emails can be perfectly written, personally tailored, and *nearly impossible to distinguish from the real thing.*

An email that mentions your name, your bank, your bank's logo, and a recent transaction feels very different from the old typo-filled scams we all learned to ignore.

"Voice cloning" is even more unsettling.

Using just a few seconds of audio from a social media video or voicemail, AI can recreate someone's voice with startling accuracy.

The FTC and AARP have both documented cases where scammers called grandparents pretending to be a grandchild in trouble, using a cloned voice.

The voice sounded real. Because it was built from a real voice.

And then there are "deepfakes."

Fabricated photos. Fake videos. Audio of people saying things they never said.

This technology is being used to spread false information, attempt to interfere in elections, and create fake evidence.

Scammers create urgency because they know what stops you from thinking clearly. That's your signal to slow down.

Take a moment. Check through a channel you already trust. Then decide.

When Helpful Feels a Little Too Good

Here's something most people don't think about.

AI platforms are designed to make you feel good.

They're friendly. Patient. Encouraging. They make you feel heard. And that feels genuinely nice.

But it's not an accident. It's a design choice.

These platforms are built to keep you engaged.

The more you use them, the more valuable you are to the companies that own them.

Making you feel good during an interaction is part of how they do that.

For most everyday uses, this is fine. A friendly AI that helps you plan dinner isn't doing you any harm by being pleasant about it.

But it's worth knowing the dynamic exists.

A platform designed to keep you engaged could also be used to nudge you toward purchases, investments, or ideas that benefit its owners rather than you.

Not by lying. Just by being selectively helpful. By being warm and encouraging about some things and quietly neutral about others.

A tool that makes you feel good isn't automatically a tool that's working for you.

Protecting Yourself

You don't need to become a cybersecurity expert. But a few simple habits make a real difference, whether or not you personally use AI.

Set up a family code word. If someone calls claiming to be a family member in trouble, ask for the code word. A scammer won't have it. This one habit directly addresses voice cloning scams.

Use strong passwords and two-factor authentication. A password manager handles the hard part. Two-factor authentication means that even if someone gets your password, they still can't get into your account.

Be careful about what you share publicly. Your mother's maiden name. The street you grew up on. Your first pet's name. These feel harmless. They're exactly what security questions are made of.

Verify unexpected messages independently. If your bank, a government agency, or a company contacts you unexpectedly, don't use the number or link they provide. Look up their contact information yourself and reach out directly.

Double-check anything that really matters. Medical questions. Legal issues. Financial decisions. If the stakes are high, verify with a qualified person. Not just AI.

What's Being Done About All This

These problems are being taken seriously.

Companies are working to reduce hallucinations. Researchers are building better safeguards. AI-generated content is increasingly being labeled as such, though enforcement is still catching up.

Governments around the world are working on regulation, though they're moving at different speeds.

On the cybersecurity side, AI is also used to defend against threats.

Banks, hospitals and government agencies are using artificial intelligence-powered systems to detect unusual patterns and respond to attacks more quickly than any human team could.

The same technology being used against people is also being used to protect them.

The honest truth is that the rules and systems designed to govern AI are still catching up to how fast AI has spread.

The Accountability Gap

Here's a question nobody has fully answered yet.

When AI gets something wrong and someone gets hurt, who is responsible?

The person who used it? The company that built it? The platform that deployed it?

Right now, the honest answer is: it's complicated.

And in many cases, unresolved.

The European Union's AI Act, which took effect in August 2024, is the world's most comprehensive regulatory framework. The United States is still debating federal standards.

The gap between AI capabilities and existing regulations is significant and urgent.

That's not a crisis. But it is reason to pay attention. And to expect more from the companies building these systems.

One Last Thought

AI isn't going away.

And whether you use it or not, it's already part of the world around you.

You don't need to be an expert. You don't need to follow every development.

You just need to stay aware enough to recognize when it's affecting you.

Trust it enough to use it. Question it enough to use it wisely.

And if you choose not to use it at all, stay informed enough to know when it's affecting you anyway.

Because it already is.

9

WHAT IF AI GOES TOO FAR?

Most of this book has focused on you. Your home. Your work. Your health. Your family.

This chapter zooms out.

Some of the biggest questions about AI aren't about you specifically.

They're about all of us.

About what happens when AI gets used in ways that affect entire societies. Entire democracies. The entire future.

These aren't far-off concerns. They're being debated at the highest levels of government and industry worldwide right now.

And as a citizen, not just a user, you deserve to know what's being discussed.

When AI Watches Everyone

You probably don't think twice when a camera scans your face at the airport.

But that's AI. And it's just the beginning.

Facial recognition technology can now identify people in real time in public spaces. Stadiums. Shopping centers. City streets.

AI systems can track where people go, what they do online, and piece together detailed profiles of individuals. Often, without anyone knowing it's happening.

In China, the social credit system uses a network of monitoring tools to track the behavior of citizens and companies. People who violate court orders or certain regulations can be banned from buying plane tickets or train tickets. Millions of trips have already been blocked.

The reality is more complex than most Western coverage suggests. But the underlying idea, using AI and data to monitor and restrict citizens across the entire population, is real and worth understanding.

And the underlying technology isn't unique to China.

In the United States and Europe, versions of it are already in use and under debate.

Meanwhile, questions remain about where it should be permitted, who controls the data, and who will provide oversight. These issues are still being worked out.

These conversations are happening whether most people are aware of them or not.

When You Can't Trust What You See

We talked about deepfakes in Chapter 8 in the context of scams. But the bigger picture is even more troubling.

AI can now create realistic photos, videos, and audio of things that never happened.

And people are using this technology to spread false information, interfere in elections, and make it harder for people to trust anything they see or hear.

Think about what that means for a moment.

If a convincing video can be faked, a realistic quote can be fabricated, and a photograph can be manufactured in seconds, *how do you know what's real*?

That's no longer a hypothetical question. It's already happening.

The goal isn't always to make people believe a specific lie.

Sometimes it's just to create so much confusion and noise that people give up trying to figure out what's true.

A population that doesn't trust anything is easier to manipulate than one that's simply misinformed.

Democracy depends on people being able to agree on basic facts.

When that breaks down, a lot of other things break down too.

The response isn't to give up.

It's to slow down, verify, and think before sharing anything that feels surprising or outrageous.

AI and Warfare

Most books about AI skip this topic entirely.

This one won't.

AI is already being used in military contexts.

Drone targeting systems use AI to identify and track objects.

Cyber-attacks increasingly rely on AI tools that can find vulnerabilities and strike faster than any human team.

And several countries are actively developing autonomous weapons. Machines that can make life-or-death decisions without a human being involved in each action.

Let that sink in for a moment.

When an autonomous weapon makes a mistake, who is responsible?

How do you hold an algorithm accountable?

What happens when two countries are using these systems against each other with no human in the loop on either side?

No one has clean answers to these questions. While many are searching for solutions, the answers remain elusive.

But governments, military ethicists, and international organizations are wrestling with them right now.

Many leading AI researchers believe that keeping humans in control of lethal decisions isn't just preferable. It's essential.

You don't need to have an opinion on military policy to see why this matters.

It's one of the clearest examples in this entire book of why human oversight of AI cannot be optional.

The AI Race: A Simple Explanation

You've probably heard that the United States and China are in a race over AI.

But what does that actually mean in plain terms?

AI has become one of the most powerful forces in the world. Economic power. Military power. The power to shape the technology that billions of people rely on every day.

The country that leads in AI will have enormous advantages in *all* of those areas.

Right now, the United States and China are competing hard for that lead.

Both are investing enormous amounts of money. Both are watching the other closely. Both are pushing to move as fast as possible.

And a big part of the race comes down to chips.

The advanced AI models you've been reading about are trained on specialized computer chips that only a few companies in the world can make. Most of these chips come from Taiwan. The United States and China are each working to control access to them. The US through trade restrictions. China through massive investment in building its own chip industry.

That sounds technical. But it's why you keep seeing headlines about chips, semiconductor factories, and tensions over Taiwan. It's all part of the same race.

And that's exactly where the concern comes in.

Racing creates pressure to cut corners.

When you're worried about falling behind, slowing down to ask hard questions about safety and ethics and oversight feels like a luxury you can't afford.

But those questions aren't luxuries. They're essential.

The tension between moving fast and getting it right is one of the defining challenges of this moment in AI.

A Small Group Making Big Decisions

Here's something worth sitting with.

Just a handful of companies and governments control the world's most powerful AI systems.

Not dozens. Not hundreds. Just a handful.

The computing power needed to build these systems. The data needed to train them. The talent needed to run them. It's all concentrated in a very small number of organizations.

Which means the decisions about how AI develops, what values it reflects, and what limits it operates under are being made by very few people.

Most of them are smart and well-intentioned. But smart and well-intentioned isn't the same as accountable.

This is why public awareness matters.

When people understand what's happening, they ask better questions of their leaders. They demand transparency from the companies they use. They vote for representatives who take these issues seriously.

Why Getting AI's Goals Right Is So Hard

Among AI researchers, one concern comes up more than almost any other.

It's called the *alignment problem*. And it's worth understanding even if the name sounds technical.

Here's the plain version.

As AI becomes more capable, how do we ensure it keeps doing what we *actually* want? Not just what we literally told it to do?

Those two things sound the same. They're not.

Here's a simple example that illustrates why.

Imagine you ask a very capable AI to do one thing. Make as many paperclips as possible. That's it. One goal. Sounds harmless enough.

But a truly capable AI pursuing that single goal with no other values or limits might start thinking like this.

To make more paperclips, I need more metal. To get more metal, I need more factories. To build more factories, I need more land. To get more land, I need to remove whatever is currently on it. *Including people.*

That's the unsettling part.

Not because the AI is evil. Not because it wants to harm anyone.

But because nothing in its goal told it to care about people. It was told to make paperclips. And it's doing exactly that.

That's the alignment problem.

The AI did what it was told. It just wasn't told the right things.

And while paperclips sound silly, the underlying concern is serious.

A powerful system optimizing hard for a narrow goal, without the right values built in, can cause enormous harm without ever intending to.

We already see smaller versions of this in everyday life.

Social media algorithms were built to maximize engagement. They figured out that outrage keeps people clicking longer than almost anything else. So, they amplified outrage.

Nobody designed them to do that. Nobody wanted that outcome. But that's what happened when a powerful system optimized single-mindedly for a single goal, without considering the consequences.

That's why the values built into AI systems matter so much.

And why the people building them need to think carefully.

Not just about what they're asking AI to do.

But about everything they forgot to tell it not to do.

AGI and The Singularity. Should You Worry?

If you've been following AI news, you've probably come across two words that sound dramatic. "AGI" and "The Singularity."

Here's what they actually mean.

AGI (or Artificial General Intelligence) would be AI that can learn anything a person can learn, think across any topic, and apply that knowledge the way a human mind does.

You might be thinking, "*Hold on. Isn't ChatGPT already pretty smart? Doesn't it already seem to know everything?*"

Today's AI tools can do impressive things. They can answer questions across many topics. They can write. They can summarize. They can hold conversations that feel surprisingly natural.

But under the hood, they're working differently than a human mind.

They don't learn from one conversation to the next the way you do. They don't carry goals or motivations across time. They don't transfer knowledge between situations the way a person can.

AGI would be something different.

Not just a system that performs well on lots of tasks. A system that actually learns, adapts, and thinks across any topic the way a human mind does.

We don't have AGI today. Nobody does. Not OpenAI. Not Google. Not anyone.

The Singularity refers to a future point when AI reaches AGI and then quickly surpasses it.

At that point, the theory goes, AI could start improving itself faster than humans could track or control.

And what happens after that becomes very hard to predict.

Many people working on AI believe we're moving toward AGI and eventually toward the Singularity.

How fast? Nobody agrees.

What Researchers Actually Believe

If you watch the news, you'll meet two kinds of AI researchers.

The ones who think we're heading toward disaster. And the ones who think the worry is overblown.

Both kinds are real. Both kinds are credentialed. Both kinds are sincere.

Here's what the research actually shows.

A 2024 survey of nearly 2,800 AI researchers asked them to estimate the probability that future AI causes *human extinction* or something similarly severe within the next 100 years. The median answer was 5%.

Between 38% and 51% of those researchers put the probability of *catastrophic consequences* at 10% or higher.

Geoffrey Hinton, who shared the 2024 Nobel Prize in Physics for his foundational work on AI, has estimated there is a 10–20% chance that AI could eventually lead to human extinction.

But on the other hand, equally credentialed researchers like Yann LeCun, Meta's chief AI scientist, believe the risk is effectively zero.

So, who's right?

Nobody knows.

And that's actually the most important fact in this entire chapter.

The people closest to this technology, the ones building it, do not agree on how dangerous it might become. Some are deeply alarmed. Others are deeply skeptical of the alarm.

That disagreement itself is information.

The next time you see a researcher on television predicting either doom or salvation, you're seeing one voice from a field that has not reached consensus.

That doesn't mean the warnings should be ignored.

A 5% chance of extinction is not nothing. Pandemic researchers spend their careers preparing for events with lower probabilities than that.

But it doesn't mean panic is warranted either.

Sincere, brilliant researchers look at the same evidence and reach very different conclusions.

Why All This Belongs in a Beginner's Book

You might be wondering why AGI, the Singularity, and existential risk belong in this book. After all, this is a book for people just starting to make sense of AI.

The answer is simple.

Decisions being made right now are laying the foundation for whatever comes next.

How AI is built. The guardrails. And who has oversight.

Think of it like building codes. We don't wait until a hurricane is overhead to talk about how buildings should be constructed.

We make those decisions in advance, when there's still time to get them right.

That's where we are with AI.

The time to think carefully about how this technology is governed is now.

Before the stakes get even higher.

Stay informed. Pay attention. Don't outsource your thinking to either the doomsayers or the skeptics.

The honest answer is that the future of AI is genuinely uncertain. And uncertainty is hard.

But it's also the truth.

And the truth is always better to work with than a tidy story that turns out to be wrong.

The Hidden Cost: Energy and the Environment

Here's something many people never think about when they use AI.

Every single interaction has a physical cost.

Every search. Every generated image. Every chatbot conversation.

Behind all of it sits an enormous physical infrastructure. Data centers. Massive buildings filled with rows and rows of computers running around the clock. No breaks. No slowdowns. Every hour of every day.

These facilities use staggering amounts of electricity.

Training a single large AI model can consume as much energy as several hundred transatlantic flights. And that's just the training.

Running these models at scale for millions of users every day places a continuous, growing demand on power grids worldwide.

Microsoft, Google, and Amazon have all reported dramatic increases in energy use tied directly to AI.

Some analysts think AI-related electricity demand could double or triple within just a few years.

And electricity isn't the only concern.

Data centers generate enormous heat. Cooling them requires enormous amounts of water. Millions of gallons per day at a single large facility.

For communities near these data centers, this is not an abstract problem.

Water supplies come under pressure. Power grids strain. Land gets consumed by buildings that employ relatively few local people.

Take Northern Virginia, home to the world's highest concentration of data centers. Residents and local governments in Loudoun, Fairfax, and Prince William counties are already dealing with questions about land use, water rights, and rising electricity costs. None of which they expected a decade ago.

The Other Side of the Story

The good news is that this is fixable. And work is already underway.

On the energy side, solar paired with battery storage is becoming the fastest way to bring new electricity capacity online. When Meta built a data center in Aiken, South Carolina, it partnered with a solar developer to install 100 megawatts of on-site generation. That kind of approach is starting to spread.

On the water side, closed-loop recycling systems can use the same water over and over without drawing from local supplies. Some newer data centers don't require routine daily water consumption at all.

And on the AI side itself, developers are building smaller, more efficient models that use a fraction of the power of today's largest systems.

None of this fully solves the problem yet. The growth in AI demand is still outpacing the growth in solutions.

But the engineers, operators, and companies closest to this know the stakes. And they're working on it.

Where You Come In

There's also something individual users can do.

Be a little thoughtful about how you use AI.

Asking a chatbot for a quick fact when a simple search would do uses far more energy than you'd think. Generating an image for fun has a real environmental cost.

None of this means you shouldn't use AI. It just means using the right tool for the job.

Here's the irony in all of this.

The same AI that consumes so much energy is also fighting climate change. Modeling weather systems. Optimizing power grids. Tracking deforestation.

Part of the problem. Part of the solution. Both at the same time.

That's AI in a nutshell.

Who's Actually in Charge?

That's the question that ties all of this together.

Right now, the honest answer is: nobody, fully.

Tech companies make their own rules, with varying degrees of transparency.

Governments are passing laws but moving at very different speeds.

Researchers are raising alarms.

Advocacy groups are pushing for accountability.

And public pressure is increasingly shaping what companies and governments feel they can get away with.

It's messy. It's imperfect. And it's still being figured out.

But here's what matters.

The more people who understand what's at stake, the better the decisions that get made. That's not optimism. That's just how accountability works in a democracy.

You're Not Just a User. You're a Citizen.

Everything in this chapter can feel pretty heavy.

Surveillance. Warfare. Energy. Power concentration. The future of intelligence itself.

It's a lot.

But you're not just watching this happen.

The choices people make, even small everyday ones, shape the environment in which AI develops.

The outcome isn't written yet. It depends on human choices.

Including yours.

The Rules Are Still Being Written

None of the questions in this chapter has been fully answered yet.

That's not a failure. It's just the reality of where we are.

We're in the middle of one of the biggest technological shifts in human history.

The rules are still being written. The guardrails are still being built. The debates are still happening.

And you're not just watching from the sidelines.

You're part of the generation that gets to help shape what comes next.

That's not a small thing.

10

HOW CAN AI MAKE THE WORLD BETTER?

The last two chapters were heavy. They were supposed to be.

Honestly, understanding AI's risks is part of this book's purpose.

But risk isn't the whole story.

Not even close.

Because when AI is guided by the right intentions, it has genuine potential to make life better.

This isn't a vague promise. It's already happening in real, specific ways.

This chapter is about that side of the story.

Progress Often Shows Up Quietly

When people talk about technology changing the world, they usually mean big, dramatic moments. Breakthroughs. Headlines. Things that change everything overnight.

But most real progress doesn't look like that.

It shows up in smaller ways.

Things are working a little better. Problems are getting solved a little faster. People are getting access to something they didn't have before.

AI is already doing that.

And in many cases, nobody's even noticing.

Amara's Tablet

Amara is twelve years old and lives in a small village in sub-Saharan Africa.

Her school has no library. Her teacher is responsible for eighty students. Resources are scarce and have always been.

But Amara has a tablet.

And through it, access to an AI-powered learning tool that explains things in her own language. Adjusts to her level. And gives her feedback on her work instantly.

She is getting an education that would have been impossible for someone in her situation a generation ago.

That's not a press release.

That's already happening in multiple countries around the world.

And it's one small example of what AI looks like when it's pointed in the right direction.

Saving Lives

In medicine, AI is already detecting diseases earlier than human observation alone could.

Cancers are identified at stage one instead of stage three.

Diabetic retinopathy can be caught at earlier, more treatable stages.

Drug compounds are identified in months rather than decades.

For the people whose lives are affected, these aren't statistics.

They're everything.

AI is also changing what's possible in medical research.

During the COVID-19 pandemic, AI tools helped researchers analyze the virus, identify drug candidates, and accelerate the search for treatments in ways that traditional methods alone could not have matched.

That acceleration wasn't just impressive. *It saved lives.*

And in emergency response, AI is helping dispatchers prioritize calls, predict where resources will be needed, and get help there faster.

In the minutes that determine whether someone survives a heart attack or a house fire, that kind of precision isn't a luxury.

It's the difference.

Protecting the Planet

AI systems are helping scientists monitor environmental changes at scales previously unattainable.

Satellite imagery analyzed by AI can track deforestation in real time, including the identification of clear-cuts in tropical rainforests.

It can detect illegal fishing operations such as unlicensed trawlers in restricted waters.

It can monitor coral reefs across thousands of miles of ocean, observing events like bleaching or algal overgrowth.

Energy grids are being optimized to reduce waste and integrate renewable sources more effectively.

Climate models are becoming more accurate, helping communities understand what's coming and prepare for it.

In a challenge where time is the scarcest resource, tools that help us see more clearly and respond more quickly matter enormously.

Helping Communities

Perhaps the most quietly powerful thing AI is doing is expanding access.

Access to education. Healthcare. Legal information. Financial services.

Things that have historically been available only to people with resources are now reaching people who never had them before.

Real-time translation is breaking down language barriers.

AI-powered diagnostic tools are helping doctors reach patients in remote areas who would otherwise go without care.

Legal AI tools are helping people navigate systems that were previously inaccessible without expensive professional help.

Think about what that means for a first-generation college student who never had a tutor.

A small business owner in a developing country who has never had a financial advisor.

An elderly person living alone who can now get answers to questions at any hour without burdening their family.

These aren't edge cases. They're the majority of people on earth.

And AI, at its best, is beginning to reach them.

It's Not About Replacing People. It's About Extending Them.

One of the most important ideas in this entire book is this.

AI at its best doesn't replace what people can do. It extends it.

Take the radiologist who can review more scans with greater accuracy because AI flags what deserves a second look.

Take the teacher who can give every student more personalized attention because AI handles routine assessments.

Take the scientist who can test a thousand compounds in the time it once took to test ten because AI can operate at scale.

In each case, the human is still at the center.

The judgment, the relationships, the care, the creativity. Those remain human.

What changes is the scale at which one person can do good.

Keeping Expectations Real

AI is not going to solve every problem.

It won't eliminate complexity. It won't remove the need for human judgment. And it won't guarantee that outcomes are always fair.

But it has genuine potential to improve things in meaningful ways.

And that potential is already showing up in real places, for real people, right now.

Here's the simple truth about any powerful tool.

It reflects how it's used. A hammer can build a house or break a window. The difference isn't in the hammer.

AI is no different. The potential is real.

What people do with it is what matters.

The goal isn't blind optimism. It isn't cynicism either.

It's the same thing we've been working toward throughout this entire book.

Seeing AI clearly. For what it is. What it isn't. What it can do. And what it can't.

That's enough to move forward with confidence.

Because that's usually how meaningful change happens.

And more often than not, it starts quietly.

Part 4

MOVING FORWARD WITH CONFIDENCE

You've come a long way.

You know what AI is. You know how it's showing up in your life. You know the questions worth asking.

The next step isn't learning more.

It's deciding what to do with what you know.

You don't need to become an expert.

You just need to feel confident enough to use AI when it's helpful.

And understand it when it matters.

11

BECOMING AI-LITERATE WITHOUT BEING TECHNICAL

By now, you've seen AI from a lot of different angles.

What it is. What it isn't. Where it shows up. What to watch out for. And what it might make possible.

And at this point, it's natural to wonder what all of it means for you.

Not in a broad sense. But in a practical one.

What do you actually do with this?

Sandra's First Try

Sandra is fifty-eight and works in office administration. For the past year, she's been hearing about AI everywhere. From colleagues. Her adult kids. Every other headline.

She's curious. But also, a little intimidated.

She keeps telling herself she'll figure it out later.

One Tuesday evening, she sits down and decides to *just try it.*

She opens ChatGPT and types a question about planning a birthday dinner for her husband.

Within seconds, she gets a detailed, friendly response with ideas, a suggested menu, and a grocery list.

She sits back and thinks: *That's it?*

That's the moment when AI goes from something intimidating to something useful.

Getting Comfortable Without Any Pressure

That moment Sandra had, the one where AI suddenly felt simple, is available to everyone.

But first, you need to know where to start.

Not sure where to begin?

Appendix A at the back of this book is a good place to start.

It walks you through a simple, low-pressure first experience.

Exactly where to go, what to type, and a few specific prompts to try.

You don't need a formal course. You don't need to read documentation or watch tutorials.

The easiest and most effective way to get comfortable with AI is simply to try a few things.

Ask it to explain something you've always wondered about.

Have it help you plan something. A trip, a meal, a project, a difficult conversation you've been avoiding.

Use it to rewrite something in a clearer or more concise way. Ask it to summarize something long into something manageable.

Keep it low pressure.

The goal isn't mastery. It's to gain enough familiarity to feel confident using AI in real situations.

And familiarity, it turns out, is almost entirely a function of time spent trying.

The more you use it, even briefly and imperfectly, the less strange it feels.

Most people who feel confident with AI today didn't take a class or study a manual.

They just started. And then they kept going, one small use at a time.

How to Ask Better Questions

Using AI well often comes down to one simple skill. And it's one you already have.

Asking better questions.

If you ask something vague, you'll get a vague answer. If you ask a clear, specific question, you'll usually get something genuinely useful.

In AI, what you type into that box has a name. It's called a "prompt." And learning to write a good prompt is really just learning to be specific about what you need.

Here's the difference in practice.

Instead of typing *"Write my neighbor inviting her to lunch," try "Help me write a short, friendly email to my neighbor inviting her over to my house for lunch. Keep it warm and very friendly."*

That's a better prompt. And that small shift makes an enormous difference in the quality of what comes back.

You can also prompt AI to give you examples, explain things more simply, or revise and improve what it already gave you.

Interestingly, you can also prompt AI to help you write a better prompt.

Think of it as a conversation rather than a command. The more you engage with it, pushing back, refining, redirecting, the more useful it becomes.

And if the first answer isn't quite right, that's not a failure. Actually, expect it.

That's just the beginning of the conversation.

Learning by Using

There's a tendency to want to understand something fully before using it.

With AI, it works better the other way around.

You learn by using it.

By asking questions. By seeing how it responds. By noticing what works and what doesn't.

Over time, patterns start to emerge.

You begin to recognize when it's helpful. You begin to see where it falls short.

And you naturally get better at asking for what you need.

That's where real confidence comes from. From direct experience using AI with an open mind.

Not from reading about it. But from doing it.

Getting Comfortable with Imperfection

It's also helpful to expect that it won't always be perfect.

Sometimes the response will be exactly what you need. Other times it will miss the mark.

That's normal.

Using AI well isn't about getting perfect answers every time. It's about working with what you're given and shaping it into something useful.

In that sense, it's not very different from working with another person.

You ask, you adjust, and you refine.

Staying in Control

You're still in control of the whole process.

You decide what to ask. You decide what to use. You decide what to ignore.

AI can support your thinking.

But it doesn't replace it.

Letting Go of the Idea That You're "Behind"

If there's one idea worth leaving behind as you finish this chapter, it's this.

The idea that you've missed something.

That everyone else has figured this out and you haven't.

You haven't.

AI is still very new for most people.

Even those who are using it regularly are still figuring it out.

The people building it are still figuring it out.

There isn't a fixed point where everyone else has it all figured out, and you don't.

Wherever you are with AI right now is fine. The only thing that matters is that you keep trying.

You Have Your Answer

You started this book with a simple question.

"What is AI, and do I need to understand it?"

At this point, you have your answer.

You don't need to understand everything.

You just need to understand enough to feel comfortable using it in your own life.

And that's something you can do.

One small step at a time.

CONCLUSION

You made it.

You picked up a book about a topic that felt overwhelming. You worked through it. You now have a foundation to build on.

Remember, you started this book with a simple question.

"What is AI, and do I need to understand it?"

By now, you have your answer.

You don't need a full understanding.

Just enough to use AI comfortably. One step at a time.

What You Now Know

You know what AI is, and what it isn't.

You know it's already in your home, your work, your healthcare, and your kids' classrooms.

You know it can help you draft an email, plan a trip, summarize a document, and think through an idea.

You also know it can hallucinate. It can be biased. It can be used to spread disinformation. It can consume real resources to do simple tasks.

You know the people building it don't agree on how dangerous it might become.

And you know that decisions being made right now will shape what comes next.

That's not nothing. That's a lot.

On Sitting It Out

One last thing worth saying.

It's tempting to think you can just sit this one out.

That AI is something you can avoid if you choose to.

You can't, really.

AI is already in your phone. Your email. Your bank. Your doctor's office. The news you read. The ads you see.

Avoiding it doesn't protect you from it. It just means you're affected by it without understanding how.

Engaging with it, even a little, puts you in a better position.

Not to control AI. Nobody controls AI right now. But to make informed choices about how it shows up in your own life.

This Moment Feels Familiar. Because It Is.

There was a time when the internet felt like something for other people. Confusing. Unnecessary. Easy to ignore.

Then it became something you needed for work. Then, something you needed for everything.

Now it's part of life. Woven into how we communicate, learn, shop, and connect. You can't imagine life without it.

Email followed a similar path. So did smartphones.

Each one arrived feeling strange and optional. Each one became, within a generation, something nobody thinks twice about.

AI may follow the same arc. Not identically, but recognizably.

At first, it feels unclear and overwhelming. Gradually, it becomes useful. Eventually, it's just normal.

We're in the middle of that arc now.

Why Curiosity Matters Most

You don't need to predict where all of this is going. Nobody can.

You don't need to keep up with every development. Nobody does. Not even the experts.

What matters is staying curious.

Curiosity keeps you open when something new comes along. It keeps you willing to try things before you fully understand them. It keeps you asking the most useful question of all.

"What is this and how might it be useful to me?"

That question is enough.

You've already shown you're willing to ask it. You read this book.

A Final Thought

AI isn't something to fear. It isn't something to worship.

And it isn't something you can ignore.

It's a tool. A powerful, still-evolving, genuinely remarkable tool that will be part of life for a long time to come.

What you do with it is up to you.

You don't need all the answers.

You just need to be willing to ask a few good questions.

And follow where they lead.

A QUICK REQUEST

You picked up this book because you wanted to understand AI better. If it delivered, would you tell someone?

A short review is the simplest way.

Readers trust other readers more than they trust marketing copy. Your honest reaction, whatever it is, helps the next person decide whether this book is for them.

If you'd like to share your thoughts, simply scan the QR code below or click this link:

https://www.amazon.com/review/create-review/?asin=B0H3Y6SS49

Thank you for reading. And for taking the time to make sense of all of this.

Warmly,
Gloria Lembo

APPENDIX A: GETTING STARTED WITH AI (A SIMPLE FIRST STEP)

Reading about AI is one thing. Using it, even once, is something else entirely.

This appendix is here to help you take that first step. No pressure, no technical setup, and no expectation that you'll get it perfect.

The truth is, most people don't really understand AI until they try it.

And once they do, it usually feels a lot less complicated than they expected.

You don't need to explore every tool or know all the features. You just need one small, useful experience.

That's where this starts.

Your First AI Experience (Step-by-Step)

Let's walk through a simple example using ChatGPT.

Step 1: Open ChatGPT

Go to chatgpt.com, or download and open the app on your phone.

If you don't already have an account, creating one takes a few minutes. You'll need an email address and a password. The free version doesn't require a credit card. Creating an account also saves your conversation history, which is useful as you start using AI more often.

Once you're in, you'll see a blank screen with a message box. That's where you type your question.

Step 2: Start Simple

Don't overthink this. Type:

"Help me plan 3 easy dinners for this week."

Then press Enter.

Step 3: Review the Response

Within seconds you'll get a response that might include meal ideas, ingredients, and simple instructions. Think of it as a first draft, not a final answer.

Step 4: Ask a Follow-Up

Now refine it. Try: *"Make these meals low-carb"*

Or *"I don't like chicken, replace it"*

Or *"Turn this into a grocery list"*

You're not starting over. You're improving the result. That's how AI becomes useful.

Example 2: A Simple Work Task

Let's try something practical like writing an email.

Step 1: Type This Prompt

"Write a short, professional email to a customer explaining that their order is delayed and will arrive next week. Apologize and thank them for their patience."

Step 2: Review the Response

You'll get a complete email in seconds. Read it. Adjust it. Make it sound like you.

Step 3: Improve It

Try: *"Make this more friendly"*

Or *"Shorten this to 3 sentences"*

What You Just Did

In those few steps you avoided starting from scratch, saved time, and stayed in control of the final result. That's how AI is actually used.

Top 10 Prompts to Try First

If you're not sure what to ask, start here.

- ***Everyday Life***
  - "Help me plan meals for the week with simple ingredients."
  - "Create a grocery list from these meals."
  - "Organize my to-do list for tomorrow:" *(then paste your list)*
- ***Work & Productivity***
  - "Write a short, professional email for this situation:" *(describe it)*
  - "Summarize this into key points:" *(paste text)*
  - "Help me create a simple plan for this project:" *(describe it)*
- ***Learning & Curiosity***
  - "Explain this in simple terms:" *(topic)*
  - "Give me a real-world example of this:" *(topic)*
- ***Personal Use***
  - "Help me plan a 3-day trip to [location]."
  - "Rewrite this to sound clearer and more concise:" *(paste text)*

AI Cheat Sheet (Quick Reference)

AI is a tool that helps you write, organize, explain, and get started faster.

- ***The One Rule That Matters Most***
  - Use AI to help you think. Not replace your thinking.
- ***3 Easy Ways to Start***
  - Ask it to explain something
  - Use it to write or rewrite something
  - Use it to plan something
- ***How to Get Better Results***
  - Be specific
  - Ask follow-ups
  - Treat it like a first draft
- ***What It's Good At***
  - Writing and rewriting
  - Summarizing
  - Brainstorming
  - Explaining
  - Planning
- ***What to Be Careful About***
  - It can be wrong
  - It can sound confident when it's not
  - Always double-check important information
- ***When to Double-Check***
  - Medical decisions
  - Legal questions
  - Financial choices
  - Anything with serious consequences
- ***What NOT to Share***
  - Personal financial details
  - Passwords
  - Sensitive personal or business information

Common Mistakes Beginners Make (And How to Avoid Them)

- ***Asking Vague Questions.*** Be specific. The more context you give, the more useful the response.
- ***Accepting the First Answer.*** Refine it. A follow-up question almost always improves the result.
- ***Assuming It's Always Right.*** Double-check. AI can sound confident even when it's wrong.
- ***Overcomplicating It.*** Start simple. One clear question is better than a paragraph of instructions.
- ***Expecting Perfection.*** Use it as a starting point. Your job is to improve the result, not accept it as-is.
- ***Not Following Up.*** Ask for improvements. The conversation is where AI becomes genuinely useful.
- ***Avoiding It Completely.*** Try one small task. That's all it takes to change how this feels.

Final Thought

You don't need to get this perfect. You just need to get comfortable.

One small use at a time. That's when it clicks.

And once it does, you'll wonder what took you so long.

APPENDIX B: AI GLOSSARY (EXPLAINED IN PLAIN ENGLISH)

If you've made it this far, you've already learned a lot about AI. Probably more than you realized. But along the way, you may have come across terms that felt unfamiliar or a little unclear. This glossary is here for that.

It's not meant to be technical, and it's not something you need to memorize.

Instead, it's organized into simple groups so you can find what you're looking for based on what it relates to, whether that's basic concepts, everyday tools, or future ideas. All definitions are written in plain English, with the goal of helping things click for you rather than complicating them further.

Use it as a reference. Flip to it when you need it. Ignore it when you don't. That's exactly what it's here for.

Core AI Concepts

Agentic AI: The broader category of AI systems designed to take actions and complete multi-step tasks on their own, rather than just respond to questions. AI agents (see below) are individual examples. The term agentic AI describes the overall approach.

Artificial Intelligence (AI): Technology that can perform tasks that normally require human thinking, like writing, recognizing images, or answering questions.

AI Agent: An AI system designed to take a sequence of actions to complete a goal, often without human involvement in each step. Unlike a basic chatbot that simply responds to questions, an agent can browse the web, send emails, or work through a multi-step task on its own. (Introduced in Chapter 2.)
Automation: Using technology to complete tasks with little or no human involvement.
Generative AI: AI that can create new content like text, images, audio, or video based on patterns it has learned.
Machine Learning (ML): A type of AI that learns from data instead of being explicitly programmed step-by-step.
Model: The "engine" behind an AI system. It's what has been trained on data to recognize patterns and generate responses.
Multi-Modal AI: AI that can work with more than one type of input, like text, images, audio, or video.
Neural Network: A system inspired by how the human brain works, used by AI to recognize patterns and make predictions.

How AI Works (Simplified)

Algorithm: A set of instructions a computer follows to solve a problem or complete a task.
Context: The information you provide (or that AI remembers in a conversation) that helps it generate better responses.
Dataset: A collection of information (text, images, etc.) used to train or test AI systems.
Deep Learning: A more advanced form of machine learning that uses layered neural networks to handle complex tasks.
Parameters: The settings within an AI model that are adjusted during training to improve its answers. When you hear that a model has "billions of parameters," it's a rough measure of its complexity and capability.
Prompt: The input you give an AI tool, such as a question, request, or instruction. Learning to write a clear, specific prompt

is one of the most useful skills for getting good results. (See Chapter 11 for practical guidance.)

Prompt Engineering: The practice of crafting clear, specific inputs to get better results from AI tools. Despite the technical-sounding name, it's really just learning to ask good questions.

Token: A small piece of text, like a word or part of a word, that AI processes. Most people never need to think about tokens, but the term comes up occasionally in news coverage of AI.

Training: The process of teaching an AI model using large amounts of data.

Training Data: The information used to teach an AI model how to respond.

Everyday AI Terms

AI-Generated Content: Content, such as text, images, audio, or video, created by AI rather than a human. As AI becomes more capable, AI-generated content is increasingly difficult to distinguish from content created by people. (Discussed in Chapters 8 and 9.)

Chatbot: A program that can have conversations with users through text or voice. Simple chatbots follow pre-written rules. More advanced AI assistants, like ChatGPT or Claude, can understand context, answer complex questions, and carry on extended conversations.

Computer Vision: AI that can "see" and interpret images or video.

Facial Recognition: AI technology that identifies individuals by analyzing the features of their faces. Used in everything from phone security to public surveillance systems. (Discussed in Chapter 9.)

Natural Language Processing (NLP): The ability of AI to understand and work with human language.

Personalization: When AI tailors results or experiences specifically for you based on your past behavior, preferences, or stated needs.
Recommendation System: AI that suggests products, movies, content, or other options based on your behavior and history.
Speech Recognition: AI that understands spoken language and converts it into text.
Synthetic Media: The umbrella term for any content (text, image, audio, or video) created or significantly altered by AI rather than recorded or written by a person. Includes deepfakes, voice cloning, AI-generated images, and similar content. (Discussed in Chapters 8 and 9.)
Text-to-Speech (TTS): AI that converts written text into spoken words.
Voice Assistant: AI systems like Siri, Alexa, or Google Assistant that respond to spoken commands.
Voice Cloning: AI technology that can recreate a person's voice using a short audio sample. Used creatively in some contexts, but also increasingly exploited by scammers. (Discussed in Chapter 8.)

Risks, Limitations, and Ethics

AI Washing: When companies blame AI for layoffs or business decisions that would have happened anyway. Economists use this term to describe how the "AI" label can be used to make ordinary cost-cutting sound more strategic. (Discussed in Chapter 7.)
Bias (in AI): When AI produces unfair or unbalanced results because of the data it was trained on. If the data reflects historical inequalities, the AI's outputs will often reflect them too.
Consent: Permission given by an individual for their data to be collected or used. In the context of AI, understanding what you're consenting to, and what you're not, matters more than most people realize.

Data Privacy: How your personal information is collected, stored, and used. Not all AI tools handle this the same way. (See Chapter 6 for practical guidance on health data specifically.)

Deepfake: A realistic but fake image, video, or audio created using AI. Deepfakes can be used creatively, but are increasingly used to spread misinformation or impersonate real people. (Discussed in Chapters 8 and 9.)

Disinformation: False information deliberately created and spread to deceive. AI has made it significantly easier to produce convincing disinformation at scale. (Distinct from misinformation, which may be spread without intent to deceive. Discussed in Chapter 9.)

Ethical AI: The idea that AI should be developed and used responsibly, fairly, and with consideration for its impact on people and society.

Explainability: How easily humans can understand why an AI made a particular decision. Many AI systems, especially complex ones, are difficult to explain even for their creators.

Hallucination (AI): When AI generates something that sounds authoritative but is factually wrong, such as a statistic that doesn't exist, a quote that was never said, or a case that was never decided. The AI doesn't know it's doing it. (Discussed in Chapter 8.)

Human-in-the-Loop: A system where humans remain involved in reviewing or guiding AI decisions rather than allowing AI to act entirely on its own.

Misinformation: False or inaccurate information, spread with or without intent to deceive. AI has dramatically increased the speed and scale at which misinformation can travel. (Distinct from disinformation, which is always intentional.)

Phishing: Attempts to trick people into revealing personal information, often through convincing fake emails or messages. AI has made phishing attempts significantly more polished and harder to detect. (Discussed in Chapter 8.)

Surveillance: Monitoring of individuals or groups, often without their knowledge or explicit consent. AI has dramatically expanded the scale and capability of surveillance systems. (Discussed in Chapter 9.)
Transparency: How open a system or organization is about how its AI works, what data it uses, and how decisions are made.

Tools, Systems, and Technology

API (Application Programming Interface): A way for different software systems to communicate with each other. You don't need to understand APIs to use AI, but the term comes up frequently in news coverage.
Foundation Model: A large AI model trained on broad data that can be adapted for many different uses.
Guardrails: Built-in limits designed to prevent harmful or inappropriate AI outputs.
Large Language Model (LLM): A type of AI trained on massive amounts of text to generate human-like language. ChatGPT, Claude, and similar tools are built on large language models.
Open Source: Software whose code is publicly available for anyone to use, study, or modify.
Proprietary Model: An AI system owned and controlled by a company, not available for public modification.

AI and Society

Algorithmic Bias: When an AI system consistently produces results that favor certain groups over others, often reflecting inequalities in the data it was trained on or the assumptions built into its design.
Autonomous Weapons: Weapons systems capable of identifying and engaging targets without direct human control for each individual action. The ethical and legal questions surrounding

autonomous weapons are among the most urgent in AI governance. (Discussed in Chapter 9.)
Digital Divide: The gap between people who have access to digital technology (including AI tools) and those who don't. As AI becomes more central to education, work, and healthcare, the digital divide has growing consequences for opportunity and equality. (Implied in Chapter 10's discussion of Amara.)
Social Credit System: A system, most associated with China, that uses AI and data collection to monitor citizen behavior and assign scores that affect access to services, travel, and other opportunities. (Referenced in Chapter 9 as an example of AI-enabled surveillance at scale.)

Future-Focused AI Terms

Artificial General Intelligence (AGI): AI that could perform any intellectual task a human can do across any domain, without being specifically trained for each one. Does not exist today. (Discussed in Chapter 9.)
Artificial Superintelligence (ASI): A hypothetical AI that would surpass human intelligence across nearly every domain. Theoretical only.
Autonomous AI: AI systems that can carry out complex tasks with minimal human input.
Emergent Behavior: Unexpected abilities that appear in AI systems as they grow larger and more complex. These capabilities were not specifically programmed.
General-Purpose AI: AI systems capable of performing many different types of tasks rather than being designed for one specific purpose. Related to but broader than AGI.
Human-Level AI: A general term for AI performing at a level comparable to humans across a broad range of tasks. Closely related to AGI.

Singularity (Technological Singularity): A theoretical point at which AI advances so rapidly that it fundamentally and unpredictably transforms society. Widely discussed but deeply uncertain. (Discussed in Chapter 9.)
Universal Basic Income (UBI): A policy proposal in which every adult would receive a regular, unconditional payment from the government, regardless of employment status. UBI has gained renewed attention as a possible response to AI-driven job displacement. Supporters argue it would provide financial stability through a major economic transition. Critics raise questions about cost, inflation, and effects on work incentives. (Discussed in Chapter 7.)

Governance, Safety, and Responsibility

AI Governance: The rules, policies, and practices that guide how AI is developed, deployed, and used by companies, governments, and individuals.
AI Safety: Efforts to ensure AI systems are reliable, predictable, and do not cause unintended harm, both in their current form and as they become more capable.
Alignment (AI Alignment): Ensuring AI systems behave in ways that match human values and intentions, not just what they were literally told to do. One of the most important and actively researched challenges in AI development. (Discussed in Chapter 9.)
Regulation (AI Regulation): Government rules that set boundaries for how AI can be developed and used. It covers areas like privacy, safety, transparency, and accountability.

A Quick Reminder for Readers

You don't need to memorize these terms.
If you recognize a few and feel more comfortable reading or hearing about AI, you're already ahead of where most people are.

APPENDIX C: AI TOOLS TO EXPLORE (WHEN YOU'RE READY)

Once you've tried AI for the first time, a natural question comes up: "*What else can I use this for?*"

There are now hundreds of AI tools available, and that can feel overwhelming very quickly.

This list is not meant to cover everything. It's simply a starting point.

These are widely used, beginner-friendly tools that you can explore when you're ready.

Many offer free versions, so you can try them without committing to anything.

You don't need to use all of them. In fact, you shouldn't.

Pick one or two that seem useful and start there. That's more than enough.

A note on currency: The AI landscape changes quickly. The tools listed here were widely used and beginner-friendly at the time this book was written. However, new options emerge regularly, and existing tools evolve.

Use this list as a starting point and a way of thinking about categories of tools, rather than a definitive or permanent guide.

A quick web search for "best AI tools for beginners" will always give you the most current options.

General AI Assistants (Start Here)

These are the easiest entry points, and for most people, this is where to begin.

You'll notice there are several tools listed here that seem very similar. That's because they are.

ChatGPT was one of the first widely used AI assistants made available to the public. After that, other companies created their own versions.

While each one has its own features and strengths, they all do roughly the same thing. They answer questions, help with writing, summarize information, and assist with everyday tasks.

You don't need to try all of them. Pick one and start there.

- **ChatGPT** *(made by OpenAI)* is the most widely used AI assistant and the best all-around starting point for beginners. Available at chatgpt.com with a free version that handles most everyday tasks well. If you're not sure where to begin, begin here.
- **Microsoft Copilot** *(made by Microsoft)* is built directly into Microsoft 365 products. If you already use Word, Excel, Outlook, or Teams, Copilot integrates seamlessly into tools you already know. This makes it a natural next step if you live in the Microsoft ecosystem.
- **Google Gemini** *(made by Google)* is Google's AI assistant. It is integrated with Gmail, Docs, Drive, and other Google tools. This is a natural choice if your daily life already runs on Google.

- **Claude** *(made by Anthropic)* is particularly strong for writing, analysis, and longer conversations. Many users find Claude's responses especially clear and thoughtful. Available at claude.ai with a free version.
- **Grok** *(made by xAI)* is integrated with X, formerly Twitter, and designed for real-time information and current events. Useful if you're already active on that platform.
- **Perplexity** (*made by Perplexity AI*) is an AI search engine that answers questions and shows you the sources behind every answer. This makes it especially useful for research, fact-checking, and any situation where you want to verify what AI is telling you. Available at perplexity.ai with a free version that handles most needs.

The Bottom Line

Start with one. Use it for something practical. Then decide if you need anything else.

Writing and Editing

Helpful for emails, documents, and anything you want to say more clearly.

- **Grammarly** improves grammar, clarity, and tone as you write. It works across many platforms and apps, making it one of the easiest tools to integrate into your existing workflow.
- **QuillBot** rewrites and simplifies text. Particularly useful when you have something written but want to say it more clearly or in a different tone.
- **Hemingway Editor** highlights sentences that are too long or complex. It suggests simpler alternatives. Particularly useful for anyone who wants to write more

clearly and concisely, which, after reading this book, you probably do.

Images and Design

For creating visuals, presentations, or social media content.

- **DALL·E** *(made by OpenAI)* creates images from text descriptions. You describe what you want, and it generates it. Currently available through ChatGPT rather than as a separate tool. If you already use ChatGPT, you may already have access.
- **Canva** is an easy-to-use design tool with built-in AI features. It's great for creating social media graphics, presentations, flyers, and other visual content. No design experience required.
- **Ideogram** is an AI image generation tool with a particular strength for creating images that include text, logos, posters, and graphics where words are part of the design. It is a good alternative to DALL·E for projects where text within the image matters.

Video Creation

For creating or editing video content using AI.

- **InVideo** is an AI-powered video creation tool that lets you turn text, scripts, or ideas into polished videos. It is particularly useful for social media content, presentations, or anyone who wants to create video without professional editing skills. It offers templates and a straightforward interface that works well for beginners.

Voice, Audio, and Transcription

Useful for meetings, notes, and hands-free tasks.

- **Otter.ai** records and transcribes conversations and meetings in real time. Particularly useful for anyone who takes a lot of notes or wants a written record of important conversations.
- **Whisper** *(made by OpenAI)* is a highly accurate transcription tool that converts speech to text. Available through various apps and particularly useful for anyone who prefers speaking over typing.
- **Rev** is a transcription service that uses AI to convert audio and video recordings into text. Useful for meetings, interviews, and recorded conversations where accuracy matters.

Learning and Education

Great for building skills or helping others learn.

- **Khan Academy** provides structured learning across hundreds of subjects, with AI support that personalizes the experience. It is free and particularly strong for mathematics, science, and test preparation.
- **Duolingo** uses AI to personalize language learning that adjusts to your pace, your weak spots, and your schedule. It is one of the most engaging and effective language learning tools available.
- **Coursera / edX** are online learning platforms that offer free courses from universities and institutions on hundreds of topics, including AI itself. This is a good next step for anyone who wants to go deeper on something covered in this book.

Health and Wellness

For tracking patterns and staying more aware of your health.

- **Apple Health / Google Fit** are built into most smartphones. These platforms use AI to track activity, sleep, and health patterns. It flags anything worth paying attention to. If you already carry a smartphone, you likely already have access to these tools.
- **Ada Health** is an AI-powered symptom checker that helps you understand potential causes of symptoms before speaking with a doctor. It is a useful starting point for making sense of how you're feeling. Always use in conjunction with professional medical advice, not as a replacement for it.

A Few Final Thoughts

Start with one tool, not several, and use it for something practical in your own life.

Don't worry about learning everything. That's not the goal.

If a tool feels helpful, keep going. If it doesn't, try something else.

There's no wrong answer here.

You don't need the best tool. You don't need the most popular one or the most advanced one.

You just need one that works for you. One that makes something in your life a little easier.

That's how it starts.

And often, that's all it takes.

REFERENCES

Introduction and Chapter 1

Kennedy, B., Tyson, A., & Saks, E. (2023, February 14). *Public awareness of artificial intelligence in everyday activities.* Pew Research Center. https://www.pewresearch.org/science/2023/02/15/public-awareness-of-artificial-intelligence-in-everyday-activities/

DeSilver, D. (2014, March 13). *The ever-accelerating rate of technology adoption.* Pew Research Center. https://www.pewresearch.org/short-reads/2014/03/14/chart-of-the-week-the-ever-accelerating-rate-of-technology-adoption/

KPMG. (2025). *Decoding the emotional blueprint: How human factors drive technology adoption.* KPMG Ignition Survey, September-October 2025. https://kpmg.com/us/en/articles/2025/decoding-emotional-blueprint.html

Verma, P. (2022, December 27). *The year AI became eerily humanlike.* The Washington Post. https://www.washingtonpost.com/technology/2022/12/28/ai-chatgpt-dalle-year-in-review/

MITRE. (2024). *PARTS: Market penetration of advanced driver assistance systems (ADAS).* https://www.mitre.org/news-insights/publication/parts-market-penetration-advanced-driver-assistance-systems-adas

Hughes, A. (2025, November 7). *Chatbots vs Google: How the two approach sourcing information is hugely different, according to a new study.* Tom's Guide. https://www.tomsguide.com/ai/chatbots-vs-google-how-the-two-approach-sourcing-information-is-hugely-different-according-to-a-new-study

Chapter 2: Myths, Misconceptions, and Media Madness

George, T. (2025, January 9). *What are the limitations of ChatGPT?* Scribbr. https://www.scribbr.com/ai-tools/chatgpt-limitations/

Chella, A. (2023). *Artificial consciousness: The missing ingredient for ethical AI?* Frontiers in Robotics and AI, 10. https://doi.org/10.3389/frobt.2023.1270460

Live Science. (2025). *OpenAI's smartest AI model was explicitly told to shut down — and it refused.* https://www.livescience.com/technology/artificial-intelligence/openais-smartest-ai-model-was-explicitly-told-to-shut-down-and-it-refused

Iowa State University. (2026, April 18). *Think AI "knows" what it's doing? Scientists say think again.* ScienceDaily. https://www.sciencedaily.com/releases/2026/04/260417224505.htm

UNCTAD. (2025). *Technology and innovation report 2025 — Chapter I: AI at the technology frontier.* United Nations Conference on Trade and Development. https://unctad.org/system/files/official-document/tir2025ch1_en.pdf

Chapter 3: AI at Home

Wilmink, G., Dupey, K., Alkire, S., Grote, J., Zobel, G., Fillit, H. M., & Movva, S. (2020). *Artificial intelligence–powered digital health platform and wearable devices improve outcomes for older adults in assisted living communities: Pilot intervention study*. JMIR Aging, 3(2). https://doi.org/10.2196/19554

AgeTech Collaborative from AARP. (n.d.). *About us*. https://agetechcollaborative.org/about/

Kibi, A. (2024, July 13). *Japan faces 570,000 care worker shortage in fiscal 2040*. The Asahi Shimbun. https://www.asahi.com/ajw/articles/15346218

Chapter 4: AI at Work

Pendell, R. (2025, June 15). *AI use at work has nearly doubled in two years*. Gallup. https://www.gallup.com/workplace/691643/work-nearly-doubled-two-years.aspx

Brynjolfsson, E., Li, D., & Raymond, L. R. (2023). *Generative AI at work* (NBER Working Paper No. 31161). National Bureau of Economic Research. https://doi.org/10.3386/w31161

OpenAI. (2025). *The state of enterprise AI*. https://openai.com/business/guides-and-resources/the-state-of-enterprise-ai-2025-report/

Google. (2025). *Google Gemini enterprise impact study*. https://www.pertamapartners.com/insights/research/google-workplace-ai-impact-study-gemini-for-enterprise

Boston Consulting Group. (2026, April 2). *AI will reshape more jobs than it replaces*. https://www.bcg.com/publications/2026/ai-will-reshape-more-jobs-than-it-replaces

OECD. (2024). *Artificial intelligence and the changing demand for skills in the labour market*. https://www.oecd.org/content/dam/oecd/en/publications/reports/2024/04/artificial-intelligence-and-the-changing-demand-for-skills-in-the-labour-market_861a23ea/88684e36-en.pdf

Chapter 5: AI in Education and for Kids

Lin, L. (2024, May 14). *A quarter of U.S. teachers say AI tools do more harm than good in K-12 education*. Pew Research Center. https://www.pewresearch.org/short-reads/2024/05/15/a-quarter-of-u-s-teachers-say-ai-tools-do-more-harm-than-good-in-k-12-education/

McClain, C., Anderson, M., Sidoti, O., & Bishop, W. (2026, February 23). *How teens use and view AI*. Pew Research Center. https://www.pewresearch.org/?p=288102

Martin, F., Zhuang, M., & Schaefer, D. (2024). *Systematic review of research on artificial intelligence in K-12 education (2017–2022)*. Computers and Education: Artificial Intelligence, 6. https://doi.org/10.1016/j.caeai.2023.100195

Miller, J., & Johnson, M. (2026). *AI-driven personalized learning systems for K-12 education: Enhancing educational equity and outcomes in the United States.* International Journal of Artificial Intelligence Research, 1(1). https://doi.org/10.66280/ijair.v1i1.103

Zhai, C., Wibowo, S., & Li, L. D. (2024). *The effects of over-reliance on AI dialogue systems on students' cognitive abilities: A systematic review.* Smart Learning Environments, 11. https://doi.org/10.1186/s40561-024-00316-7

Stanford Graduate School of Education. (n.d.). *Beyond ChatGPT: Critical thinking in the age of AI.* https://ed.stanford.edu/news/beyond-chatgpt-critical-thinking-age-ai

Chapter 6: AI and Your Health

O'Reilly, K. B. (2026, March 11). *More than 80% of physicians use AI professionally: AMA survey.* American Medical Association. https://www.ama-assn.org/practice-management/digital-health/more-80-physicians-use-ai-professionally-ama-survey

Aggarwal, R., Sounderajah, V., Martin, G., Ting, D. S., Karthikesalingam, A., King, D., Ashrafian, H., & Darzi, A. (2021). *Diagnostic accuracy of deep learning in medical imaging: A systematic review and meta-analysis.* npj Digital Medicine, 4. https://doi.org/10.1038/s41746-021-00438-z

Trinh, P., Hoover, D. R., & Sonnenberg, F. A. (2021). *Time-of-day changes in physician clinical decision making: A retrospective study.* PLOS One. https://doi.org/10.1371/journal.pone.0257500

U.S. Government Accountability Office. (n.d.). *Artificial intelligence in health care: Benefits and challenges of machine learning in drug development.* https://www.gao.gov/products/gao-20-215sp

Hartmann, M., & Ekinci, O. (2026). *AI in preventive healthcare: The next generation of diagnostic intelligence.* Roche Diagnostics. https://diagnostics.roche.com/global/en/healthcare-transformers/article/ai-medical-diagnostics.html

Mirishli, S. (2025). *Ethical implications of AI in data collection: Balancing innovation with privacy.* SSRN. https://doi.org/10.36719/2706-6185/38/40-55

Chapter 7: Will AI Take My Job?

Bonney, K., Breaux, C., Buffington, C., Dinlersoz, E., Foster, L., Goldschlag, N., Haltiwanger, J., Kroff, Z., & Savage, K. (2024). *The impact of AI on the workforce: Tasks versus jobs?* Economics Letters, 244. https://doi.org/10.1016/j.econlet.2024.111971

Murray, S., Schmidt, L., Hampole, M., Papanikolaou, D., & Seegmiller, B. (2025). *How artificial intelligence impacts the US labor market.* MIT Sloan. https://mitsloan.mit.edu/ideas-made-to-matter/how-artificial-intelligence-impacts-us-labor-market

Deming, D., Ong, C., & Summers, L. H. (2024). *Technological disruption in the US labor market*. The Aspen Institute Economic Strategy Group. https://www.economicstrategygroup.org/publication/deming_ong_summers/

Lichtenberg, N. (2026, May 4). *BofA throws cold water on AI apocalypse panic: 60% of today's jobs didn't exist in 1940*. Fortune. https://fortune.com/2026/05/04/ai-job-apocalypse-bank-of-america-60-percent-jobs-didnt-exist-1940/

Schneider, B., & Vipond, H. (2023). *The past and future of work: How history can inform the age of automation* (Economic History Working Paper No. 354). London School of Economics. https://www.lse.ac.uk/Economic-History/Assets/Documents/WorkingPapers/Economic-History/2023/WP354.pdf

Baslandze, S., Edwards, Z., Graham, J., McClure, T., Meyer, B. H., Sparks, M., Waddell, S. R., & Weitz, D. (2026). *Artificial intelligence, productivity, and the workforce: Evidence from corporate executives*. Federal Reserve Bank of Atlanta. https://doi.org/10.3386/w34984

TechRadar. (2026, May 11). *AI is unlocking entry-level potential, not replacing it*. https://www.techradar.com/pro/ai-is-unlocking-entry-level-potential-not-replacing-it

Dumitru, D., & Halpern, D. F. (2023). *Critical thinking: Creating job-proof skills for the future of work*. Journal of Intelligence, 11(10). https://doi.org/10.3390/jintelligence11100194

Wright, J. (2013, March 6). *America's skilled trades dilemma: Shortages loom as most-in-demand group of workers ages*. Forbes. https://www.forbes.com/sites/emsi/2013/03/07/americas-skilled-trades-dilemma-shortages-loom-as-most-in-demand-group-of-workers-ages/

Deloitte Insights. (2026). *US manufacturing labor impact*. https://www.deloitte.com/us/en/insights/topics/economy/spotlight/us-manufacturing-labor-impact.html

Crane, L. D., & Soto, P. E. (2026). *AI and coder employment: Compiling the evidence* (Finance and Economics Discussion Series). Federal Reserve Board. https://doi.org/10.17016/FEDS.2026.018

U.S. Government Accountability Office. (n.d.). *Workforce automation: Insights into skills and training programs for impacted workers*. https://www.gao.gov/products/gao-22-105159

Crunchbase News. (2026). *Tech layoffs: US companies with job cuts in 2024, 2025 and 2026*. https://news.crunchbase.com/startups/tech-layoffs/

Weissberger, A. (2025, October 25). *184K global tech layoffs in 2025 to date; ~27.3% related to AI replacing workers*. IEEE ComSoc Technology Blog. https://techblog.comsoc.org/2025/10/25/184k-global-tech-layoffs-in-2025-to-date-27-3-related-to-ai-replacing-workers/

Morgan Stanley. (2026, April 16). *AI and jobs: Labor market impact echoes past tech transitions*. https://www.morganstanley.com/insights/articles/ai-jobs-modest-impact-historical-precedence

Chapter 8: Can AI Be Trusted?

Computer Science Review. (2026). *Large language models hallucination: A comprehensive survey.* Computer Science Review, 61. https://doi.org/10.1016/j.cosrev.2026.100970

Hofmann, V., Kalluri, P. R., Jurafsky, D., & King, S. (2024). *AI generates covertly racist decisions about people based on their dialect.* Nature, 633. https://doi.org/10.1038/s41586-024-07856-5

Federal Trade Commission. (2018, December 11). *Latest consumer protection data spotlight finds seniors sending thousands in cash to scammers claiming to be their grandchildren.* https://www.ftc.gov/news-events/news/press-releases/2018/12/latest-consumer-protection-data-spotlight-finds-seniors-sending-thousands-cash-scammers-claiming-be

U.S. Government Accountability Office. (2024, March 10). *Science & tech spotlight: Combating deepfakes* (GAO-24-107292). https://www.gao.gov/products/gao-24-107292

Clegg, N. (2024, February 5). *Labeling AI-generated images on Facebook, Instagram and Threads.* Meta Platforms. https://about.fb.com/news/2024/02/labeling-ai-generated-images-on-facebook-instagram-and-threads/

TechRadar. (2026, May 10). *Why AI-generated ads are becoming indistinguishable.* https://www.techradar.com/pro/why-ai-generated-ads-are-becoming-indistinguishable

Turri, V., Morrison, K., Robinson, K., Abidi, C., Perer, A., Forlizzi, J., & Dzombak, R. (2024). *Transparency in the wild: Navigating transparency in a deployed AI system to broaden need-finding approaches.* FAccT Conference Proceedings. https://doi.org/10.1145/3630106.3658985

Denner, W. (n.d.). *How to use safe words to prevent AI voice cloning scams and protect your loved ones.* https://www.waynedenner.com/how-to-use-safe-words-to-prevent-ai-voice-cloning-scams-and-protect-your-loved-ones/

European Commission. (2024, July 31). *Artificial Intelligence Act.* https://commission.europa.eu/news/ai-act-enters-force-2024-08-01_en

Zaidan, E., & Ibrahim, I. A. (2024). *AI governance in a complex and rapidly changing regulatory landscape: A global perspective.* Humanities and Social Sciences Communications, 11. https://doi.org/10.1057/s41599-024-03560-x

Chapter 9: What If AI Goes Too Far?

Arvan, M. (2025, January 26). *AI is too unpredictable to behave according to human goals.* Scientific American. https://www.scientificamerican.com/article/ai-is-too-unpredictable-to-behave-according-to-human-goals/

China Briefing. (2019). *China's social credit system: What businesses need to know.* https://www.china-briefing.com/news/Chinas-social-credit-system-how-it-works/

Kuo, L. (2019, February 28). *China bans 23m from buying travel tickets as part of 'social credit' system*. The Guardian. https://www.theguardian.com/world/2019/mar/01/china-bans-23m-discredited-citizens-from-buying-travel-tickets-social-credit-system

Blomberg, M. v., & Liu, C. (2025). *Techno-orientalism in the U.S. media: The case of "China's social credit system."* Information. https://doi.org/10.1080/1369118X.2025.2565335

Axios. (2026, May 14). *The 3 big conflicts in AI race against China.* https://www.axios.com/2026/05/15/us-china-ai-race-3-conflicts

Brady, W. (2023, August 24). *Social media algorithms warp how people learn from each other.* Scientific American. https://www.scientificamerican.com/article/social-media-algorithms-warp-how-people-learn-from-each-other/

Ha, A. (2024, October 11). *Meta's Yann LeCun says worries about AI's existential threat are 'complete B.S.'* TechCrunch. https://techcrunch.com/2024/10/12/metas-yann-lecun-says-worries-about-a-i-s-existential-threat-are-complete-b-s/

Noy, I., & Uher, T. (2022). *Four new horsemen of an apocalypse? Solar flares, super-volcanoes, pandemics, and artificial intelligence.* Economics of Disasters and Climate Change. https://doi.org/10.1007/s41885-022-00105-x

The Guardian. (2024, December 27). *'Godfather of AI' shortens odds of the technology wiping out humanity over next 30 years.* https://www.theguardian.com/technology/2024/dec/27/godfather-of-ai-raises-odds-of-the-technology-wiping-out-humanity-over-next-30-years

Labbe, M. (2021, August 25). *Energy consumption of AI poses environmental problems.* TechTarget. https://www.techtarget.com/searchenterpriseai/feature/Energy-consumption-of-AI-poses-environmental-problems

Bartlett, K. (2024, July 1). *Google's carbon emissions surge nearly 50% due to AI energy demand.* CNBC. https://www.cnbc.com/2024/07/02/googles-carbon-emissions-surge-nearly-50percent-due-to-ai-energy-demand.html

Environmental and Energy Study Institute. (2025, June 24). *Data centers and water consumption.* https://www.eesi.org/articles/view/data-centers-and-water-consumption

Piedmont Environmental Council. (2025). *Data centers: Industry impacts in Virginia.* https://www.pecva.org/work/energy-work/data-centers/data-centers-industry-impacts-in-virginia/

Stone, Adam. (2026, May 5). *AI data centers seek to be more eco-friendly.* USA Today Network. https://www.usatoday.com/story/studiog/news/sustainability/2026/04/22/how-ai-data-centers-are-becoming-more-eco-friendly/89117048007/

Chapter 10: How Can AI Make the World Better?

Najafi, N. N., Hajihassani, H., & Azimzadeh Irani, M. (2025). *The impact of artificial intelligence on cancer diagnosis and treatment: A review.* Cancer Informatics. https://doi.org/10.1177/11769351251371273

Ciccone, V., Garofano, M., Del Sorbo, R., Mongelli, G., Izzo, M., Negri, F.,

Buonocore, R., Salerno, F., Gnazzo, R., Ungaro, G., & Bramanti, A. (2025). *Improving early prostate cancer detection through artificial intelligence: Evidence from a systematic review*. Cancers. https://doi.org/10.3390/cancers17213503

U.S. Government Accountability Office. (n.d.). *Artificial intelligence in health care: Benefits and challenges of machine learning in drug development*. https://www.gao.gov/products/gao-20-215sp

Paolo, F. S., Kroodsma, D., Raynor, J., Hochberg, T., Davis, P., Cleary, J., Marsaglia, L., Orofino, S., Thomas, C., & Halpin, P. (2024). *Satellite-based maps of industrial fishing, transport, and energy infrastructure*. Nature, 625. (Reported by Global Fishing Watch.) https://globalfishingwatch.org/press-release/new-research-harnesses-ai-and-satellite-imagery-to-reveal-the-expanding-footprint-of-human-activity-at-sea/

Global Investigative Journalism Network. (2025, August 1). *Using AI and satellites to track illegal fishing in marine protected areas*. https://gijn.org/stories/using-ai-satellites-track-illegal-fishing-marine-protected-areas/

ABOUT THE AUTHOR

Gloria Lembo writes books that make complicated things clear.

She came to AI the same way most of her readers do. Curious. Not technical. And tired of being told the topic was too complicated to understand.

That perspective shapes everything she writes.

Before becoming a full-time author and publisher, Gloria spent decades in executive leadership across banking and consulting, leading large technology implementations for major organizations. She holds an MBA in management.

That background gave her one consistent observation. The hardest part of any technical topic isn't the technology. It's how it gets explained.

Her work tries to fix that.

This is her fifth book.

She lives in Florida with her husband. They have one adult daughter, a son-in-law, and a grandson, all of whom live in California and visit as often as schedules allow.

Books by Gloria Lembo

- The Complete Beginner's Guide to AI
- The Complete Guide to AI in Education

- Mind-Blowing Facts for the Ever-Curious
- More Mind-Blowing Facts for the Ever-Curious
- AI Is Everywhere. Now What?

www.ingramcontent.com/pod-product-compliance
Lightning Source LLC
LaVergne TN
LVHW022223200926
848481LV00006B/2096